Language, Biology and Cognition

Prakash Mondal

Language, Biology and Cognition

A Critical Perspective

Prakash Mondal
Department of Liberal Arts
Indian Institute of Technology Hyderabad
Sangareddy, Telangana, India

ISBN 978-3-030-23714-1   ISBN 978-3-030-23715-8 (eBook)
https://doi.org/10.1007/978-3-030-23715-8

© The Editor(s) (if applicable) and The Author(s) 2020
This work is subject to copyright. All rights are solely and exclusively licensed by the Publisher, whether the whole or part of the material is concerned, specifically the rights of translation, reprinting, reuse of illustrations, recitation, broadcasting, reproduction on microfilms or in any other physical way, and transmission or information storage and retrieval, electronic adaptation, computer software, or by similar or dissimilar methodology now known or hereafter developed.
The use of general descriptive names, registered names, trademarks, service marks, etc. in this publication does not imply, even in the absence of a specific statement, that such names are exempt from the relevant protective laws and regulations and therefore free for general use.
The publisher, the authors and the editors are safe to assume that the advice and information in this book are believed to be true and accurate at the date of publication. Neither the publisher nor the authors or the editors give a warranty, expressed or implied, with respect to the material contained herein or for any errors or omissions that may have been made. The publisher remains neutral with regard to jurisdictional claims in published maps and institutional affiliations.

Cover credit: Alamy PDNG6T; BNP Design Studio/Alamy Stock Photo

This Palgrave Macmillan imprint is published by the registered company Springer Nature Switzerland AG
The registered company address is: Gewerbestrasse 11, 6330 Cham, Switzerland

To my Khuki,
who always beams with curiosity

Preface

The biological foundations of language reflect assumptions about the way language relates to biology. Human language has a biological basis more perspicuously because human language is acquired by humans but not by any other species known on the planet. Additionally, the structure of the mental organization is significantly changed when language is acquired by humans. In this sense, language is a special property of the human cognitive system. This suggests that language in virtue of being supported by the biological substrate organizes the structure of mental organization in a dramatic way. Thus, language, biology, and linguistic cognition, the form of cognition that cannot exist minus language, appear to be connected to each other in non-trivial ways. Biological structures including our genetic materials and processes are thus supposed to afford the grounding for human language and the form of cognition it gives rise to. The role of biology can be apprehended not just in the acquisition of language but also in neural processing of language and language disorders. The goal of this book is to offer a critical perspective on the relationship between language, the form of cognition, and biology in order to see whether this can provide consequential insights into the nature of human language itself. While it seems clear that the biological foundations of language help understand language from a certain vantage point, the assumptions that peddle the theoretical and empirical conceptions of the biological basis of language are not really as convincing as they appear to be. In many cases, they are in fact unfounded on certain clear grounds. Once these grounds are articulated, it may turn out that there is far more complexity to human language than can be grasped through biological constraints and principles which have an undeniable impact on the emergence of language as a capacity, though. This presupposes that the role of biology is limited to certain aspects or dimensions of language that are independent of, and perhaps also segregated from, other facets of human language such as its representational properties (which make up cognitive resources) and also the logical complexities of language. If this is indeed the case, as this book argues at length, the overenthusiasm associated with theories and frameworks of the biological basis of language is misplaced and hence needs to be contained.

As a matter of fact, with the rise of biolinguistics as a field of inquiry marrying linguistics with biology we seem to have come closer to a deep understanding of language and linguistic cognition. It promises to offer an understanding of the cognitive system of language known as 'the faculty of language' as a biologically instantiated and constrained system. Thus, the logical properties attributed to language or to the operations on linguistic constructions within the language faculty are actually biological properties described at some level of abstraction. There is no case of derivation or reduction of language to biological structures. Although the Minimalist model of the language faculty bolsters, and is also supposed to facilitate, this inquiry, this is by no means restricted to those adhering to the Minimalist model of language. Hence, on the other hand investigations into the biological basis of language in many other quarters of neuroscience, linguistics, and psychology proceed on the assumption that the nature of linguistic structures can be traced to certain biological structures and/or processes. Thus, aspects of cognitive structures for language are supposed to be rooted in biological structures and/or processes. Compelling though the logic of such investigations may seem, the fallacies in the reasoning deployed in such investigations may not become immediately apparent. This is exactly the role this book is intended to play. In a nutshell, this book argues that relations of neurobiological and genetic instantiation between language and the underlying biological substrate are, to all intents and purposes, irrelevant to understanding the fabric of language and linguistic cognition. Crucially, this book aims to offer an antidote to the current thinking embracing 'biologism' in linguistic sciences.

I've attempted to cover as much of the interdisciplinary territory on language–biology relations as has been possible, hoping to present in the book only the most representative cases of what have been made subject to critical scrutiny. If anything has been missed, the fault lies with me. In fact, readers are encouraged to find out linkages that have been missed. The critique articulated in the book first fleshes out the fallacies in current thinking on language–biology relations and then proposes a subtly different way of looking at the nature and form of language. Readers are asked to assiduously pass through this transition in order that they can relish the ideas to come ahead.

I invite linguists of all brands, philosophers of mind, psychologists, and biologists, especially neuroscientists, to take what they think they can imbibe from the book. Besides, anyone serious about understanding language and how it has got something to do with biology is also welcome. I wish I could have written this book for more audiences, but any group that I may have excluded would recognize that I would then have ended up writing a different book altogether!

Hyderabad, India Prakash Mondal
March 2019

Acknowledgements

I thank the three anonymous reviewers of this book project who have provided crucially meaningful criticisms of certain arguments marshaled in the book. This has resulted in the refinement of certain lines of reasoning, especially deployed in the critique part of the book. Thanks also go to the audiences consisting of scholars in linguistics, neuroscience, and the philosophy of mind who have provided helpful feedback to the central ideas in this book presented during talks at various universities. I'm especially indebted to Cathy Scott, my editor at Palgrave Macmillan, who has placed unswerving faith in the project and has also maintained her patience in making sure that the book project comes to fruition. This is especially noteworthy when the project proposal was going through rounds of review.

Finally, I wish to thank all those students who have listened to me for years but have never refrained from asking simple questions on anything I've proposed which later on have turned out to be not that simple.

Contents

1	**Introduction**	1
	1.1 On the Notions of Language vis-a-vis Biology	3
	1.2 Linguistic Cognition and the Underlying Biological Substrate	10
	References	38
2	**Biological Foundations of Linguistic Cognition**	43
	2.1 Genetic Foundations of Language and Cognition	51
	2.2 Neurobiological Foundations of Language and Cognition	65
	2.2.1 Brain Imaging Studies	66
	2.2.2 Lesion Studies, Language Disorders, and Other Neurological Cases	73
	2.3 How May Linguistic Cognition Ride on Neurobiology?	84
	2.4 Summary	89
	References	90

3	**Cognition from Language or Language from Cognition?**	97
	3.1 Language from Cognition	99
	3.2 Cognition from Language	115
	3.3 Language–Cognition Synergy	134
	3.4 Summary	137
	References	138
4	**Linguistic Structures as Cognitive Structures**	141
	4.1 The Cognitive Constitution of Linguistic Structures	142
	4.1.1 Variable Binding	143
	4.1.2 Quantifiers	158
	4.1.3 Complex Predicates	184
	4.1.4 Word Order	202
	4.1.5 Two Types of Grammar	216
	4.2 Summary	228
	References	229
5	**Conclusion**	233
	References	238
Index		239

List of Figures

Fig. 3.1	The head vs. non-head organization of a simple motor action	124
Fig. 3.2	One motor action embedded within another motor action	125
Fig. 3.3	The X-bar organization of a motor action	125
Fig. 4.1	The path between the binder and the variable in (55–56)	144
Fig. 4.2	The path between the binder and the variable in (59–60) (NP = noun phrase, QNP = quantificational noun phrase, VP = verb phrase, S = sentence)	145
Fig. 4.3	Non-convex and convex spaces	171
Fig. 4.4	The convex region in the connected space of 'all'	173
Fig. 4.5	The convex region in the connected space of 'some'	175
Fig. 4.6	The convex and non-convex regions in 'no'	177
Fig. 4.7	The convex and non-convex regions in 'only'	178
Fig. 4.8	The convex and non-convex regions in 'many'	182
Fig. 4.9	The circular mapping path from intra-event mapping to event linking	200
Fig. 4.10	The circular mapping path from intra-event mapping to event linking	200

Fig. 4.11 The derivational and representational types of grammatical representation (TP = Tense Phrase, *v*P = Light Verb Phrase, VP = Verb Phrase, DP = Determiner Phrase, PP = Prepositional Phrase. The arrows in the tree diagram under the derivational type of grammars represent displacement operations, and *f1* ... *f3* under the representational type represent different functions that map elements of dependency tree to phrasal constituents) 219

1

Introduction

The intrinsic nature of language is such that it permits languages to be acquired by both children and adults, to be represented in the brain/mind and also to be used by human beings who achieve linguistic competence within certain biological constraints that modulate or govern the learning and processing of language. Thus, the biological grounding of language makes it possible for the abstract system of language to be instantiated in human beings who then press into service the language capacity to accomplish various actions such as thinking, communicating, conceiving which partake of, exploit and interface with a host of cognitive capacities. From this perspective, it seems reasonable to believe that the biological basis of the language capacity offers insights into the nature of cognitive capacities such as reasoning, learning, memorization, sensory-perceptual conceptualization only insofar as the language capacity is supposed to make transparent many aspects of cognitive structures and mechanisms that constitute the cognitive substrate. The character of the human mind seems to be visible from the biological lens of language once we assume that the nature and form of cognitive structures and mechanisms can be deduced from the biologically grounded connection between the language capacity and other cognitive capacities.

© The Author(s) 2020
P. Mondal, *Language, Biology and Cognition*,
https://doi.org/10.1007/978-3-030-23715-8_1

Neurological and genetic studies on the relations between the language capacity and other cognitive capacities such as vision, memory, non-visual sensory perception, learning, reasoning, motor abilities, emotion can thus be believed to shed light on the texture of our cognitive makeup. It may be noted that such studies are supported by the supposition that the biological basis of language itself meshes well with studies on the relationship between language and cognitive capacities and/or processes. From this, it appears that biology acts as a kind of bridge that relates language to cognition given the presupposition that the path from language to cognition cannot be traversed directly.

The aim of this book is to show that the transitions from biology to language and then from language to cognition are not only hard but also invalid on many grounds. This may eventually show that language is far more closely coupled to cognition than is usually thought. Thus the present book will argue that biology cannot be the bridge that relates language to cognition or connects cognition to language because language *in itself* constitutes the system that links to cognition directly without requiring any immediate grounding relation that biology may establish. That is, the purpose of this book is to demonstrate that cognition is not transparent to biology, contrary to mainstream thinking on the relationship between biology and cognition. If cognition is transparent to something, it must be language. This is not, however, to deny that cognition—or language, for that matter—has a grounding in biology, or that cognition is a biological function that modulates many physiological-chemical processes inside organisms (Lyon 2006; Tommasi et al. 2009). In fact, many interactions with the environment that constrain learning, perception, memory, reasoning, action, etc. are instantiated in the physiological and biochemical processes within our bodies. But the crucial point to be noted is that the physical instantiation of cognition in our biological substrate is not sufficient for an understanding of what cognition is, or of how it really works. In other words, just because we understand how X is instantiated in Y, we may not come to understand X. From the fact that we understand how X is instantiated in Y it does not follow that we also understand X. Plus the direction of explanation may not simply go from the physical instantiation of cognition in our biology to an understanding of cognition itself.

The argument to be advanced in the present book is that the direction of explanation instead goes from cognition *as revealed through language* to biology. Thus, biology cannot be the appropriate medium that can give us a purchase on the problem of understanding the intrinsic nature and form of cognition with a special reference to natural language. Language being the *sine qua non* of cognitive capabilities or faculties can be the right link which can take us inside the interior space of our cognition. In this connection, it is also of particular concern to emphasize that the transition from biology to cognition is barely understood, while the path from language to cognition is in a much better shape for an exploration of the issue of how language–cognition relations can help penetrate the realm of cognition by bypassing the instantiation relation with reference to biology.

1.1 On the Notions of Language vis-a-vis Biology

It is necessary to appreciate that language is a very tricky word: some people use it to mean the faculty of language (the system of grammar that is instantiated as a modular system as part of the human mind), or linguistic competence (the competence in language X, for instance), or as a collective term for languages (as in 'Bengali is my [first] language'). It is thus worthwhile to note that these different entities do not all relate to biology in the same way. When one thinks of language (in the sense of the language faculty) as a critical component of human cognition, language is conceived of as a mental organ instantiated in the cognitive substrate just as the stomach or liver is instantiated in the human digestive system (Chomsky 2000). This conception of language invariably inherits a relationship with biology in the sense that language is now a component of the neural architecture whose properties can be discovered and studied only by relating to the underlying principles governing the development, maturation, and functions of the neurobiological infrastructure. Hence a unification of the cognitive sciences with the biological sciences is often sought on the grounds that many questions about the grounding or implementation of language within the biological substrate can be faithfully answered as problems in the unification become more and more tractable. Needless to say, this conception of language presupposes an integral or inherent relationship between language and biology since language is itself a biological entity on this view. Now if we turn to another conception of language under which language is thought of as linguistic competence (in a given language), a full-blown linguistic system that has been internalized by a human being is what is at issue. Language on this conception *can* be a biological entity, but this implication is not necessary, for if language equates to linguistic competence, the competence is true of a given language and may well be internalized as a cultural knowledge base from the relevant linguistic community just like rituals are internalized as a system by a human being from the surrounding cultural milieu. Although it is certainly the case that the system that constitutes linguistic competence in a certain language is psychologically represented, it does not follow that the competence can itself be a biological entity. But, of course, if the mature stabilized system that constitutes linguistic competence is thought to have passed through stages of biological growth only to be in the current state, the linguistic competence under this condition can be a biological entity. In fact, this possibility is indeed one such case that is endorsed by Generative Grammar, as the final state of the developing language faculty is characterized as linguistic competence (Chomsky 2000). In short, language recognized as the linguistic competence does not necessarily import an inherent relationship with biology.

We now focus on the third conception of language on which language is taken to be an extra-biological entity—an entity that is collectively realized as a system that can be studied and analyzed. Notably, on this conception language is instantiated not in an individual, but located in the intersubjective collective space of a linguistic community, books, codifying resources, etc. Here language is a sociocultural property whose resources are distributed over a loosely connected range of entities some of which are even inanimate or inert entities. Taken in this sense, language is grounded in the outer world where the symbolic properties of language are shared among groups of human beings with a diverse ensemble of things serving to function as props for the codification, preservation, and entrenchment of linguistic forms. Language in this sense is remotely related to the biological substrate because the containment within the individual is no longer viable as language becomes a supra-individual entity. In simpler terms, when we say that English has the rule X but not the rule Y, we are making statements about an entity which is not an individual property per se. Biology has got nothing to do with this. But note that one may attempt to draw the biological substance into the ambit of the shared knowledge of language as it sits in the intersubjective realm, by maintaining that the individual knowledge of language is sharable or transmittable only if it is biologically instantiated in a human being (see Mondal 2012). That is, the property of being transmittable is inherited from the property of individual instantiation. Nevertheless, it does not follow that the shared system in itself is a biological property or entity even though the property of sharing obtains when language as a biological entity takes the grounding within the individual. Therefore, the connection between biological instantiation and language as a cultural resource remains tenuous.

Against this backdrop, the current book aims to state something different about the notion of language to be employed with reference to its relationship with biology. The conception of language to be employed in the present context will encompass the first two notions of language (excluding the third) in ways that make it possible to distinguish between the inherently biological conception of language on the one hand and the potential biological conception of language on the other. The idea to be advanced has been largely taken from Katz and Postal (1991), Postal (2003), Mondal (2014), and Levine (2018). It needs to be clarified that these works essentially advance the claim that biological relations are ultimately irrelevant to understanding the basic texture of human language. But it is worthwhile to note in this context that Katz and Postal (1991) and also Postal (2003) essentially support a *realist* view of language which holds that languages or linguistic objects are abstractions, whereas the present work favors a non-cognitivist[1] conceptualist

---

[1] A cognitivist view usually imports an information processing functionalist perspective on the nature of the mind which a non-cognitivist view resists (see Mandler 2002).

view of language on which linguistic structures are themselves cognitive structures. In this sense, it appears that this conception accords well with the central tenets of Cognitive Linguistics (Lakoff 1987; Langacker 1987, 1999), but the crucial difference here is that the present work adopts and refines a *split ontology* of language on which language can be situated in two dimensions—the dimension of psychological or neurobiological instantiation and the dimension of symbolic abstraction. In the case of the former dimension, language or the linguistic capacity is essentially an aspect of the mind/brain and hence a system with finite resources (words, rules, constraints, etc.). But, in terms of the latter dimension, language can be projected into the realm of abstractions where infinite levels of expansion of linguistic forms, abstractions that may have no anchoring in the physical world (such as logical properties and relations found in language), universal categories, etc. can exist. These two dimensions are independent of each other and yet are somehow connected because of the mind's intentional projection of a finite system into a domain of abstractions where infinite extensions are always possible. In simpler terms, when we say that a language can have a sentence of length 10^{10}, we are not, of course, making any claim about the actual working of language on the dimension of psychological or neurobiological instantiation. Rather, we are saying that the given language allows for such an abstract generalization if a different configuration of psychological or neurobiological instantiation were available to humans with far greater cognitive resources. But this leap to this level of abstraction obtains via the mental projection. An analogy from mathematics will be apt here. For instance, even though there are psychological/neurobiological constraints on the mental processing of numbers (the length of numbers factored in) and their calculations (the exact count of numbers manipulated at a time factored in), there is nothing that prevents the human mind equipped with the knowledge of mathematics from concluding, on the basis of the fact that 10 is a natural number, that 10^{10} is also a natural number. Thus, the realm of abstractions where language operates as a pure axiomatic system is distinct from the level of psychological or neurobiological instantiation of language. That this distinction is often confounded by many scholars at the cross section of linguistics and (neuro)biology can be shown by the following textual references. What the current book calls into question is eloquently described by Salvador E. Luria (1973), one of the earliest proponents of the study of biological foundations of natural language.

> To the biologist, it makes eminent sense to think that, *as for language structures, so also for logical structures,* there exist in the brain network some patterns of connections that are genetically determined and have been selected by evolution as effective instruments for dealing with the events of life. (emphasis added, p. 141)

In a similar vein, Brown and Hagoort (2000) state the following by offering certain considerations that they think should be part of accepted common knowledge.

> ... a great deal of what we know about the structure and functioning of the language system has come from research that has essentially ignored the fact that *language is seated in the brain.* (emphasis added, p. 3)

Likewise, in thinking that a theory of language must be constrained by both what linguistic investigations reveal about the form of language and what neurolinguistic explorations tell us about the brain representation of language, Ingram (2007) states the following.

> There are many arguments, but no compelling reasons, why the organization of communication abilities in the brain should be isomorphic with any particular linguistic theory of language structure, unless, of course, *the theory in question were specifically formulated to take account of human brain structure and function.* (emphasis added, p. 41)

This presupposes that taking into consideration the brain representation of linguistic structures is a desideratum to be met for there to be an adequate theory of linguistic structures. There is another revealing passage from Bickerton (2014a) quoted below.

> ... there could obviously be two ways of describing syntax. One would provide maximal coverage of the empirical data while simultaneously achieving maximal levels of elegance, simplicity, and explanatory power. The other would adhere, as far as possible, to a literal description of what

the brain actually does in order to produce sentences. Would those two descriptions be isomorphic? Not necessarily. The first, constrained solely by the linguistic data, could legitimately use whatever devices might help it achieve its goals of simplicity, elegance, and comprehensiveness, regardless of how its solutions related to what brains actually do. *Should* those two descriptions be isomorphic? Obviously yes. To the extent that they differed, one would simply be wrong, and if they prove instead to be isomorphic, one is redundant. But which is redundant, the knowledge model or the mechanistic model? There can be no question that the former is redundant, since without the latter, there would be nothing to describe. (pp. 75–76, author's own emphasis)

In the context of a discussion on the invalidity and uselessness of the system of linguistic competence (in the model of Generative Grammar) pitted against a neural processing-based account of language, Bickerton considers it necessary to have an isomorphism between a descriptive account of the neurobiology of grammar and the descriptive system of grammar. Irrespective of whether or not the system of linguistic competence in the model of Generative Grammar is dispensable, it seems clear that Bickerton thinks that a brain-based account of language must guide the construction of a description of linguistic structures that is cognitively meaningful and explanatory.

Pulvermüller (2018) also states what in its spirit chimes with the views of the scholars cited above.

> Fortunately, recent neuroscience research has provided important insights into the specific features of human brain anatomy and function, which open new perspectives on answering the big question about the specificity of human cognition by mechanisms rooted in human neurobiology. However, to achieve this, it is necessary to spell out and understand *the mechanistic relationship between language and communication and their basis in neurobiological structure and function*. (emphasis added, p. 1)

Overall, this conveys the impression that the grounding of language in the biological substrate *must* be part of the understanding of the form of natural language. This is exactly what confounds the distinction between the abstract form of natural language and the neurobiological instantiation of language.

The claim to be advanced and defended throughout the current book is that the logical texture of the axiomatic system of language in the dimension of symbolic abstraction reveals many significant generalizations about the nature and form of linguistic cognition. The dimension of neurobiological instantiation is ultimately irrelevant to an understanding of linguistic cognition when looked through the dimension of neural instantiation. The nature and form of linguistic cognition when looked through the dimension of neural instantiation (which is natural because linguistic cognition is a facet of language taken to be a system that is psychologically or neurobiologically instantiated) is something that is pointless, primarily because linguistic cognition is rendered reference-less in lacking what it is constituted of or what it consists in. But, if linguistic cognition is looked through the dimension of symbolic abstraction, many abstract properties of language as a symbolic system render themselves amenable to the unfolding of their inherent cognitive contents. These symbolic properties of natural language(s) do not reside in biological entities or structures because they arise only when the brain extends to connect to the outer world consisting of language users, objects, events, processes, etc., thereby providing the scaffolding for such otherwise biologically meaningless symbolic patterns—which is something along the line of thinking developed in Northoff (2018). The implicit understanding here is that the dimension of neurobiological instantiation is located at a lower scale of realization than the dimension of symbolic abstraction. The central point of the claim put forward here is that reaching the realm of (linguistic) cognition through the higher scale of symbolic abstraction is to be preferred to an entry into the realm of (linguistic) cognition through the lower scale of neurobiological instantiation. It needs to be stressed that the claim here is not simply that there are many aspects of language and cognition that we cannot understand through biology, because this will be trivial on the one hand and on the other hand one can always insist that this is not due to any fault in biological studies or even in biology. Besides, from this it does not also follow that biological studies are not or cannot be explanatory. Although this is largely on the mark, this will miss the point raised here. It is certainly the case that biological studies are explanatory, and that is why we have explanations of various processes of life, diseases, instinctive behaviors, functions of biological organs, etc. But these explanatory accounts hold only within the domain of biological functions, processes, and entities. Significantly, the cognitive constituents of linguistic forms/structures are such that they are not constituted of biological substance the way bird feathers or cells, for example, are. A biological description, let alone an explanation, of the cognitive constituents of linguistic forms/structures is not actually close to the mark any more than a linguistic account of cell differentiation is biologically close to the mark. Be that as it may, biological accounts in many ways remain valid for language but only when they concern the *envelope* or contours of language as a cognitive system (in the first sense mentioned right at the beginning of this section). That is, biological accounts of the acquisition, development, and evolution of language remain restricted to the external manifestation of the cognitive system of language as a whole—its internal parts, constituents, and structures are inaccessible to biological descriptions. Nor are the internal parts, constituents, and structures of language to be captured by biological processes. Thus, for instance, no one has yet furnished (or perhaps will never be able to furnish) a biologically grounded description of the mental structuring of noun phrases. That is to say that even though biological accounts of language can reach up to the scale or dimension of psychological or neurobiological instantiation, they touch and tap *only* the outer manifestation of the capacity of language, which is after all an expression of the capacity allowing for the structural patterns in language, *but not* what resides inside the envelope. This point will be fleshed out in greater detail as we proceed. As notions of level or scale have appeared in our discussion here, there are many associated concepts structured around them that need to be clarified.

1.2 Linguistic Cognition and the Underlying Biological Substrate

Before we elaborate on the question of how language as a cognitive system can furnish entry into the realm of cognition, we think it necessary to explicate the ways in which cognition as reflected within and through language—the form of cognition that is constituted by language—can be reckoned to be instantiated in the biological substrate, and also to determine how this instantiation relation turns out to be inadequate on both logical and cognitive grounds. There are in fact two general ways in which cognition as manifested through language can be deemed to be instantiated in the biological substrate. The first route is the genetic level at which the basic biological layout of organisms along with their structures is specified, and the second route is the level of neural organization from which cognition is *naturally* supposed to emerge.[2] Now at this juncture, it is vital to recognize that the genetic level and the level of neural organization are part of a heterogeneous lattice or hierarchy of levels cutting across systems of molecular, cellular, and tissue-level organizations, neural networks, and whole neural structures/organs. Given that the genetic level and the level of neural organization are levels of description/explanation of a vastly complex scheme having interconnected part–whole relations across levels, it seems also necessary to understand why these two levels can be regarded as the crucial levels that compose the instantiation relation between biology and cognition.

First, the genetic level is, to all intents and purposes, the lowermost level in the heterogeneous lattice/hierarchy of levels appropriate to the biological instantiation of cognition, primarily because the physical level of atoms and other elementary particles that underlies the genetic level is not fine-grained enough for the expression of distinctions and descriptions that make cognition viable. In other words, the physical level underlying the genetic level is not tuned to what variegated facets of cognition constitute and exhibit. On the other hand, the level of neural organization can be considered to be the level that supports the cognitive infrastructure. Since it is situated exactly below the level at which the cognitive machinery works, there is good reason to believe that the level of neural organization can serve to *directly* implement within its ambit the instantiation relation between biology and cognition. Second, even if intermediate levels that can be thought to be

---

[2]Once cognition is shown to be related to, and ultimately anchored in, the biological substrate, it can be believed that cognition is thus naturalized.

interposed between the genetic level and the level of neural organization may also look like good candidates relevant to the biological instantiation of the cognitive machinery, these levels do not in themselves suffice to provide an adequate description of the level of cognition since these levels do not have the right (bio)logical structure that can be said to underpin cognitive structures, mechanisms, and processes. Consider, for example, levels of molecular, cellular, and tissue-level organizations. These levels do not possess the appropriate locus of description or explanation for cognitive structures, mechanisms, and processes, in that cognitive functions or capacities and mechanisms are not after all directly installed in molecules, cells, and tissues taken in isolation even if they co-compose the cognitive superstructure. Besides, the genetic level forming the basis of these intermediate levels furnishes all the necessary ingredients as well as the whole infrastructure for the emergence of these intermediate levels, thereby rendering redundant any attempt to trace the appearance of the cognitive superstructure to levels lower than the level of neural organization. To be clearer, it needs to be stressed that the level of neural organization must be construed in terms of individual neurons, neuronal assemblies, neural networks, and whole neural structures/organs. This construal of the level of neural organization in fact renders gratuitous the search for additional levels appropriate to the biological implementation of cognition.

We shall first explore how cognition as reflected through language can be instantiated in our genome. If the instantiation of cognition in the biological substrate is via the genetic level, it seems reasonable to hold that the genetic level could be said to support and thus underpin the formal structure of cognitive representations, processes, mechanisms, and interactions as mediated by language. This looks, on the face of it, like a reasonable way of having linguistic cognition implemented in the biological substrate at the lowest level within the hierarchy of descriptions relevant to the biological implementation of cognition. However, closer inspection reveals that this is illusory for various reasons. First of all, various kinds of cognitive structures that can be bootstrapped from natural language constructions cannot be couched at the genetic level—at least in terms that make them amenable to their characterization as cognitive structures. Let's take some simple and fairly well-known cases that illustrate quite well what is at stake.

(1) What do you think you could have done___ to alleviate the poverty and squalor in this town?
(2) You are the only living soul I can tell my secret to___.
(3) The more I think of her, the less I feel interest in things around myself.
(4) Never did he in his wildest dreams think that he would be awarded the coveted prize.
(5) They cannot in any event take this for granted.

All these sentences mean quite different things but what is more crucial in this context is that their structural differences are indicative of certain significant cognitive distinctions that they give rise to. Additionally, they reveal a number of disparate constructional generalizations which can be conceptualized as constraints on classes of linguistic signs and their components allowing for different degrees of granularity along the lines of Sign-Based Construction Grammar (Sag et al. 2012; see also Goldberg 2006, 2019). Significantly, these sentences uncover certain essentially related fundamental patterns of cognitive structures that can be bootstrapped or derived from natural language. The sentences in (1–2) are clear-cut cases of form-meaning divergence in natural language. That is, the relevant linguistic form appears in a place where it is not interpreted to mean what it means. Thus, for example, the *Wh*-expression 'what' is supposed to be interpreted in the gap shown in (1), and likewise, the noun phrase 'the living soul' is interpreted not where it appears but rather in the gap right after 'to.' These cases show that the cognitive structures that may underlie such linguistic structures can actually contain more than one representation of the same item which correspond, and can thus be linked, to one another. From the example in (3), one can infer that certain cognitive structures can correlate two representations by comparing and contrasting them with one another, as is evident in the presence of 'the more' and 'the less' in two different clauses. From (4) one can reasonably assume that some cognitive structures must have enriched representations that allow a whole representational structure to be constrained or modified by a single *operator-like* item. In (4) the inverted adverb 'never' scopes over and hence modifies (in fact, emphatically negates) what the clause 'he in his wildest dreams thought that he would be awarded the coveted prize' as a whole means. Finally, (5) shows that cognitive structures can have schema-like or *templatic* representations in which a few slots are fixed and a few others are variable-like in virtue of requiring something that can satisfy the appropriate role warranted. In idioms like 'take something for granted,' the slot occupied by 'something' is variable-like, while 'take __ for granted' is the fixed template. It is also noteworthy that most of these generalizations are based, at least in part, on the hierarchical arrangement of pieces of linguistic structure. For instance, the *Wh*-expression 'what' in (1) and the noun phrase 'the living soul' in (2) are separated from their gaps through a hierarchical (rather than linear) distance. That this is the case can be easily shown. If the distance is linear (calculated in terms of sequential positions), no insertion between one element and the gap matching with it can be done. But, since the distance is hierarchical, we can easily change (1) to 'What do you think she feels you could have done__…?' or (2) to 'You are the only living soul I can comfortably tell my dirty secret to__'. Likewise, if the arrangement of words was linear, the adverb 'never' in (4) would not at all scope over an entire clause because this adverb is linearly placed before only the past tense auxiliary 'did' and can thus be expected to modify the meaning of only 'did,' contrary to facts.

What these examples clearly show is that various patterns of cognitive structures that natural language offers a glimpse into are constituted by certain forms of representations which are descriptively perspicuous *only* at higher levels. The question of whether the forms of representation that patterns of cognitive structures exhibit are explanatorily perspicuous at the genetic level is a moot point. It needs to be noted that for the forms of representation exhibited by various patterns of cognitive structures to be explanatorily perspicuous at the genetic level, mechanisms of genetic transmission and other non-cultural developmental or epigenetic factors that make genetic transmission and the expression of genes viable must furnish an account of why various patterns of cognitive structures exhibit the forms they do. Also, these mechanisms will have to adequately and completely account for the character of the exact forms of representation exhibited by various patterns of cognitive structures. That is, we shall require an account of the forms of representation exhibited by various patterns of cognitive structures just like we may seek an account of the biological forms or traits found in living organisms. This would enable one to understand the ways in which the forms of representation exhibited by diverse patterns of cognitive structures can be, *largely if not exclusively*, due to the mechanisms of genetic transmission and other non-cultural developmental factors. But the deep problem is that as far as the current state of understanding is concerned the genetic level can actually be scaled up as high as the level of neural assemblies and networks. The question of how this scaling up can account for the character of the exact forms of representation exhibited by patterns of cognitive structures, as well as providing an answer to the 'why' question, is not clear. In the absence of any bridge mechanisms, the desired account eventually turns out to be a kind of quantitative description of cognitive structures that can be taken to be traits subject to different degrees of phenotypic variation within delimited populations as a function of genetic factors as well as non-genetic influences (see Bouchard 2007). The degree of phenotypic variation of a range of cognitive structures can then be calculated in a population, leading to estimates of heritability of the cognitive structures concerned. Estimates of heritability will thus leave room for the influence and interaction of non-genetic factors that contribute to the growth of cognitive structures. Even if this looks promising, this picture ends up being flaky and miserably leaves out the actual mechanisms that underlie the statistical descriptions of heritability of types of cognitive structures in humans. Nor does this tell us how certain cognitive structures in humans but not others *may*, if at all, undergo *canalization* (Waddington 1940), which as a term refers to the reliable appearance or occurrence of some trait in generations despite substantive genetic and environmental variations (biped walking in humans is one such example).

It is necessary to understand that the argument here does not merely advert to the incompleteness in our understanding of the genetic mechanisms that can account for the nature of the exact forms of representation exhibited by various patterns of cognitive structures. Rather, the argument points to the fundamental categorical mismatches that exist between the genetic level and the level of cognitive structures, mechanisms, and processes manifested in and through language. Besides, genetic coding or even genetic determination operates only at the level of protein building in cellular networks, and hence once we cross this level it becomes patently unclear how various patterns of cognitive structures or their forms of representation can be said to be structured by such protein-building processes (see Godfrey-Smith 2007; Graham and Fisher 2015). That is, it is not clear how various patterns of cognitive structures or their forms of representation can be said to be genetically coded or determined when inter-level relations are not articulated or made explicit. As emphasized just above, *only* estimates of heritability of various cognitive structures are not also sufficient. Hence the long and the short of the story is that the biological instantiation of cognition at the genetic level is (ontologically) incoherent as well as incomplete.

The other way the biological instantiation of cognition seems viable is through the neural organization. The level of neural organization seems to be the most appropriate level for explorations into the nature of biological instantiation of cognition. First of all, the neurobiological substrate provides the scaffolding for whatever cognitive capacities are constituted by, since cognitive capacities must arise from somewhere. From this perspective, the brain is the first place to look for the roots and imprints of cognitive capacities. Second, it appears to be trivially the case that what brains do is what minds are (see Pinker 1997; Ascoli 2015; Gazzaniga 2018). Although this belief has remained tantalizingly close to a thinking that adopts a reductive stance toward the derivation of mental structures, processes, and capacities from the neural machinery, there have been attempts to circumvent the difficulties that one may encounter in deriving mental structures, processes, and capacities from the neural machinery. And one of the ways of doing this is to allow the relevant derivation to pass through a series of reductive stages so that at the final stage the mental can be, possibly through a number of smooth transitions, traced to the neural at the bottom (see Churchland 1986; Churchland and Sejnowski 1992; Bickle 1998, 2003; Stoljar 2010; Silva et al. 2013).

There are a few things that need to be clarified at this stage. The derivation of mental structures, processes, and capacities from the neural machinery may not, at least in part, equate to the implementation of mental structures, processes, and capacities in the neural machinery. That is, if we have derived X from Y, it does not necessarily follow that we have actually implemented X in Y. The reason this is so is that the derivation of X from Y *may* allow for the elimination of X simply because all that we can know about X is latent in Y, whereas the implementation of X in Y does not necessarily lead to the elimination of X. This can be explicated in a better way by taking the case of program implementation in computer science. If a program is implemented in the hardware composed of electrical circuits, the program concerned does not disappear or is eliminated; rather, the hardware in which a given program is instantiated or implemented is simply ignored and the significant details about the operations running in the electrical circuits are bypassed for any description of program implementation. From this, it should be clear that the implementation of mental structures, processes, and capacities in the neural machinery need not give rise to the elimination of mental structures, processes, and capacities or, for that matter, of any description of the mental. Hence, if something abstracts away from something else, this abstraction may obtain either because the abstraction relation holds by virtue of a series of derivations or because the abstraction is simply a matter of blocking the embedding context from coming into view at the level of what is embedded.

What is, of course, important in the present context is that the argument is not simply that cognition as reflected within and through language cannot be reduced to the level of neurobiological organization. Rather, the claim is much stronger than this. The central thesis this book aims to advance is that cognition as reflected within and through language is *neither implemented in nor derived from* our neurobiological organization. One of the fundamental reasons for this argument is that there are no mechanisms at the scale of neurobiological organization that can support or underlie mental structures, processes, and capacities of language. And if no mechanisms that can support mental structures, processes, and capacities of language exist at the scale of neurobiological organization, there is no sense in which cognitive structures/

representations and capacities of language can be said to be implemented in or derived from our neurobiological organization. Although it is evident that the scale or level of cognitive systems/structures is higher than that of biological structures, it is not clear how to bring the two levels/scales into logical correspondence, especially when we are concerned about the character of linguistic cognition, or rather about the system of cognitive representations and structures manifested in language. Even if one may insist on furnishing a statement of the mechanisms of neural pathways activated during the processing of a given class of linguistic structures, this does not in any way establish that the properties of cognitive representations of the relevant linguistic structures are necessarily derived from those neural structures. Likewise, one may wish to offer a statement of the mechanisms of gene networks, gene expression, and genetic variation correlated with a variety of expressions of language disorders in order to argue for the biological implementation of cognitive properties of the linguistic representations found missing or disrupted in the given language disorders. But this does not also show that the relevant cognitive properties of the linguistic representations affected in language disorders are *constituted* by the properties of the genetic mechanisms at hand. Rather, this demonstrates only that the manifestation of linguistically structured cognition is governed by mechanisms of gene networks, gene expression, and genetic variation. Mechanisms of gene networks, gene expression, and genetic variation may govern the manifestation of cognition in the same way they govern the manifestation of the external form and structure of an organism. In this sense, governing does not equate to determination since governing reinforces certain constraints without determining how those constraints are going to be eventually instantiated through interactions with the environment outside. In view of the problems specified above, one may still hope that the deep chasm between the two modes of description can be somehow eliminated by postulating intermediate structures, mechanisms, or systems or something of that sort. However, this book will demonstrate that the chasm between the scale/level of cognitive systems/structures and that of biological structures is too wide to be eliminated or neutralized. There are many facets of the linguistic system that project deeper insights into the structural format of cognition but cannot otherwise be traced to the biological substance on the one hand, and biological structures (genetic structures and aspects of neurobiology) do not in any way attach to these facets of the linguistic system on the other.

In this connection, it must be noted that the reductive stance that traces properties of the mental to the properties of neurobiological organization has long fallen into disfavor (see for a general argument, Nagel 2012). So the idea of having a one-to-one correspondence or even identity between the mental and the neurobiological appears to be largely dogmatic and unconvincing. This is so because the mental can be assumed to be capable of being realized in multiple types of substance—whether biological or non-biological, and in addition, it is completely unclear how the mental can be derived from the level of neurobiological organization, especially when one can never be sure whether the relevant derivation is accurate or sufficient to uphold the granting of the privilege to that level from which some other level is derived. After all, the derivation could be gratuitous or even fortuitous. That this concern is not spurious may be considered by taking into account the following scenario. If one has, for example, derived a theory of human reasoning abilities and/or processes from the rules and principles of deductive and inductive logics, it could be the case that the derivation at hand builds on some statistical correlations between some regularities in actual human reasoning and the relevant principles of deductive and inductive logics that (appear to) underlie those regularities. Therefore, from this it does not unambiguously follow that all of human reasoning in various real-world circumstances can be jettisoned, just because the apparent underlying principles of deductive and inductive logics suffice to support the required derivation of all of human reasoning from those principles.

In fact, Craver (2007) and Bechtel (2008) have marshaled such arguments in order to drive home the point that reductive explanations of the kind discussed just above are doomed to be sterile and unrewarding because most neuroscientific explanations are in essence non-reductive and mechanistic-functional (but see for a different view, Egan 2017). Thus they delineate the nature of mechanistic-functional explanations in neuroscience that can serve to relate the high-level cognitive structures and capacities to the low-level neurobiological organization. The account in question presupposes that there are certain levels in the biological and cognitive sciences with respect to which a certain phenomenon can be described and ultimately explained. So a higher-level cognitive mechanism which realizes a certain phenomenon in the neurobiological substrate can exist by virtue of the organization of a number of components of that cognitive mechanism *and* the activities of these components. What this implies is that a cognitive mechanism can be fully explained by making reference to the overall organization of the component parts of the mechanism along with the activities of these component parts. If this is so, then this turns out to be a kind of whole–part relationship that holds between the mechanism and the components that enable the mechanism to realize the cognitive phenomenon in the neurobiological structures. Thus, the mechanism constitutes a level which helps realize a cognitive phenomenon at a higher level, and the components of the mechanism exist at a lower level than both the mechanism as a whole and the phenomenon realized. Such a whole–part relationship for a certain mechanism realizing a cognitive phenomenon of interest is defined with respect to the function(s) of the mechanism. That is, the organization of the component parts of the mechanism as well as the activities of these component parts must *causally* allow the mechanism to carry out the function(s) that it actually does. For example, long-term potentiation (LTP) in the hippocampus and other areas of cerebral cortex is a mechanism that realizes the cognitive phenomenon of learning in organisms including humans. This mechanism is composed of certain components which may include the activities of NDMA (N-methyl-D-aspartate) and other kinds of glutamate receptors at the cellular level which can in turn involve activities of Na+ and Ca ions at the molecular level.

It may be noted that such an account of the relationship between the cognitive level and the neurobiological level aims at achieving a decomposition of cognitive structures, capacities, and processes into their relevant parts or components that can make plausible as well as viable the ultimate correspondence with the neural structures via a series of compositional relations (see also Lycan 1987). Needless to say, the (de)compositional mapping procedure provides us with an explanatory strategy that may be supposed to present a stable descriptive bridge linking the cognitive level to the level of neurobiological organization. Now if the question of biological instantiation of cognition in the neural machinery is recast in this explanatory framework, what becomes compellingly clear is that cognition as constituted through language is neither implemented in nor derived from our neurobiological organization. The reason is simply that the mechanisms that operate at the level of neural structures are *not* the mechanisms that may underlie and support cognitive structures, capacities, and processes as reflected within and through language. Furthermore, the mechanisms that underlie and support cognitive structures, capacities, and processes shaped by language cannot be decomposed into parts or components that can be finally found at the level of neural structures or are in touch with the neural structures. It may be noted that this argument highlights two crucial aspects. The first part of the argument consists in the point that the mechanisms operating at the level of neural structures are *not* actually identical or coextensive with the mechanisms supporting cognitive structures, capacities, and processes reflected within and through language. And the second part of the argument rests on the point that *no* compositional relations hold between the mechanisms supporting cognitive structures, capacities, and processes shaped by language and the parts or components that may serve to enable the mechanisms at the cognitive level to realize the relevant cognitive structures and capacities by being in touch with the neural structures at the bottom. We shall discuss each part of this argument at length.

If we contend that the mechanisms operating at the level of neural structures are not actually identical or coextensive with the mechanisms supporting cognitive structures, capacities, and processes reflected through language, this implies that the set of mechanisms that actually operate at the level of individual neurons and/or neuronal assemblies are not included in, or do not include, the ensemble of mechanisms that (may) support cognitive structures, capacities, and processes dedicated to language. It is not hard to see that all the mechanisms at the scale of individual neurons and/or neuronal assemblies are mechanisms that actually carry out some neuronal function(s) in terms of neurotransmission. The neural activities in the retinal cells, for example, carry out complex retinal image-processing functions by way of coordinated interactions among rod and cone cells, bipolar cells, and the ganglion cells. Similarly, activities of individual neurons and/or neuronal assemblies in the brain area V4 in the visual cortex constitute neural mechanisms that execute functions of color processing of perceived objects. To give another example from the case of memory, the activities of individual neurons and/or neuronal assemblies in brain areas of the hippocampus such as CA1, CA2, and CA3 constitute mechanisms of synaptic strengthening/weakening crucial for many forms of memory consolidation/removal and learning/forgetting. Thus, what is significant for us to note is that these neurobiological mechanisms cannot in any sense *equate to* the mechanisms supporting cognitive structures, capacities, and processes of language. At best neurobiological mechanisms of such kinds can be the mechanisms that provide the neuronal basis for the performance of those mechanisms that support cognitive structures, capacities, and processes in language. This point can be appraised in a better way when one considers the role of neural activities in the Brodmann areas 44 and 45 (BA 44 and BA 45) which as patterns of interactions among neuronal networks constitute mechanisms of syntactic and semantic processing. Here, at best we can say that the neural mechanisms the neural activities in these brain regions give rise to can be deemed to be the mechanisms to which the higher-level mechanisms for cognitive structures, capacities, and processes of language may, if at all, be attributed. But in no sense can these mechanisms be regarded as the mechanisms that *in a direct manner* support cognitive structures, capacities, and processes of language. Hence the instantiation in the neurobiological substrate of the mechanisms that directly support cognitive structures, capacities, and processes of language is not viable.

Let's consider this issue in more detail. If the aforementioned conclusion were wrong, we would be able to decompose the higher-level mechanisms for cognitive structures, capacities, and processes of language into components that will eventually or ultimately turn out to be neurobiological entities and activities. Consider, for example, some of the commonly known psycholinguistic mechanisms that mediate language processing. The following examples illustrate such mechanisms well.

(6) The soldier jumped near the barbed wire fences finally collapsed.
(7) All the professors the police officers have spoken to over the phone have refused to disclose the truth.
(8) The boy broken by the news that arrived today burst into tears.
(9) We know the students of the professors who come to see me every day.
(10) There are few countries in the world which do not execute criminals.

The sentences in (5–10) serve to demarcate the domains over which some psycholinguistic mechanisms operate as constraints on possible mechanisms that can be deemed to underpin linguistic structures and representations in the mind. The sentences in (6–7) in essence show that the underlying psycholinguistic mechanism has a preference for the immediate establishment of structural relations between adjacent items. In (6) the sentence leads to processing difficulty because one assumes that the verb of the subject 'the soldier' is 'jumped,' but actually it is 'collapsed'.[3] The difficulty was evaluated in a now-famous sentence 'The horse raced past the barn fell,' which is termed a 'garden-path' sentence because of the temptation in processing to carry on with the wrong interpretation one is lulled into making only to be led down the garden path (see Frazier and Clifton 1996; Townsend and Bever 2001). It is worth noting that the particular difficulty is, at least in part, due to the overly complex noun phrase at the beginning having layers of embedding, as can be easily checked in the noun phrase 'the soldier jumped near the barbed wire fences' in example (6). In fact, this consideration applies pretty well to the sentence in (7) where the semantic dependency between the noun phrase 'all the professors' and the verb 'spoken to' is interrupted by the parts of the complex noun phrase 'all the professors the police officers have spoken to.' However, this does not straightforwardly apply to the case in (8) in which the noun phrase

---

[3]Some speakers do indeed feel that this sentence is unacceptable. Importantly, that the verb 'jump' can appear in the causative form in the context of garden-path sentences has been pointed out by Sanz (2013), although it may not be immediately obvious to some users of English.

'the boy broken by the news that arrived today' is also complex and yet does not pose a processing difficulty of the kind (6) does. So it seems that something over and above what is called 'the principle of minimal attachment' (which prefers a minimal amount of branching within phrases in the representation of linguistic structure) is at work here. In particular, this may be due to the nature of the verbs used in (6–8). The verb 'break' in (8) initiates a process that can be delimited by the object, whereas the verb 'jump' used in a causative form is not delimited by its object, for even if X's jumping is an unbounded process, X's jumping Y also remains unbounded but X's breaking Y renders the process of breaking bounded (a point made by Sanz [1996]). The other possibility is that the complex noun phrase 'the boy broken by the news that arrived today' can easily be reinterpreted as 'the boy who was broken by the news that arrived today,' which is certainly owing to the presence of 'by' in it. The sentences in (9–10) exemplify the psycholinguistic 'principle of late closure,' which bans a non-local or non-adjacent interpretation of structural relations, especially toward the end of sentences. Hence on the preferred reading 'who' in (9) refers to the 'professors' but not to the 'students.' However, this does not mean that a local attachment of an item to another item structurally close to it must have to obtain, as is evidenced by (10). Overall, it is clear that both the principle of minimal attachment and the principle of late closure are principles that favor local (as opposed to distant) structural relations among (pieces of) linguistic structures.

What these examples suggest is that the mechanisms that support cognitive structures, capacities, and processes manifested in and through language are themselves constrained by properties intrinsic to the linguistic system. That is, they reflect principles and constraints that are inherently linguistic but not neuronal. From this, we can safely state that there are stronger grounds for supposing that these psycholinguistic mechanisms are (some of) the mechanisms that support cognitive structures, capacities, and processes of language. Now if we intend to decompose these higher-level mechanisms for cognitive structures, capacities, and processes of language into their components, the components of these mechanisms (in the sense of Craver [2007] and Bechtel [2008]) will not in any way turn out to be neurobiological entities and activities.

What is at least plausible is that the components of these mechanisms may be taken to be mental strategies or subroutines that employ some attentional and/or memory resources by way of tracking and holding in a buffer (pieces of) linguistic structures. But once again the problem is that these components of the mechanisms are not neurobiological entities and/or activities. Hence, if the components of the mechanisms supporting cognitive structures, capacities, and processes of language do not 'shade into' neurobiological entities and/or activities, it is hard to see how linguistically structured cognition can be said to be instantiated in, let alone derived from, the neurobiological level of organization. The same conclusion follows when we concentrate on possible mechanisms that can support the cognitive structures uncovered from the sentences (1–5) above. Here, one may well insist that implementation imposes certain constraints on higher levels of organization that are completely dependent on the substrate. Thus, for example, computer simulations cannot run faster than a certain limit due to certain limitations imposed by the physics of electricity and the maximum speed at which information can be transferred (e.g., the speed of light would be an upper bound on some things). Likewise, one could contend that psycholinguistic preferences can definitely arise from the constraints of implementation, in that the brain cannot manipulate more than three or four dependency links at a time. But note that constraints of implementation are quite distinct from the descriptive/explanatory efficacy of implementation. This book does not dispute the former, but rather contests the latter. Even if an appeal to the constraints of implementation may account for the underlying motivations of psycholinguistic principles such as 'late closure' and 'minimal attachment', it is noteworthy that the same constraints of implementation *do* allow for the violation of these principles in other contexts where non-syntactic information plays a pivotal role and also in languages such as Spanish (Mitchell et al. 2000; Clifton 2000; Garnham 2005). Clearly, if the constraints of implementation were the actual basis of these patterns, they would serve to account for their violations as well. But that is not what these constraints do.

If the components of the mechanisms supporting cognitive structures, capacities, and processes of language cannot blend into neurobiological entities and/or activities, one may then expect that the neurobiological scale can rise to comprise the psychological level. The deeper problem is that the relevant neurobiological mechanisms cannot be 'scaled up' to form the mechanisms that (can) directly support cognitive structures, capacities, and processes of language at higher scales. As a matter of fact, this problem cannot be brushed off, or in any way circumvented, by arguing that the relevant neurobiological mechanisms operate at the level of particular brain regions, thereby enabling the mechanisms concerned to *sort of* move up across scales in order to form mechanisms that can directly support cognitive structures, capacities, and processes of language. That this line of reasoning is fraught with insuperable difficulties and hence invalid has been convincingly shown by Johnson (2009), who has argued that no mechanisms exist at the level of brain regions that carry out certain functions. The reason that he has put forward to back up his argument is that all neurobiological mechanisms operate *only* at the lower scale of individual neurons and/or neuronal assemblies. Any neurobiological mechanism that performs a certain cognitive role consists of certain neuronal parts and/or activities and thus operates at the scale of our neurobiological organization, not at the computational or psychological scale. Additionally, no cognitive functions that neurobiological mechanisms perform and, for that matter, no cognitive mechanisms can be found to occur *within* neurobiological mechanisms. Thus, if the higher-level mechanisms for cognitive structures, capacities, and processes of language cannot be broken down into parts and components that can reach down to the neurobiological scale, it follows that the mechanisms operating at the level of neural structures are not identical or coextensive with the mechanisms supporting cognitive structures, capacities, and processes in language. In this connection, it needs to be mentioned that this conclusion differs from what Johnson seems to have arrived at, for Johnson thinks that the mental may have to be ultimately traced to the scale of neurons and neuronal interactions. In the present context, it has been argued that the hard problems with the smoother transitions from the mental to the neurobiological point to something else, precisely because the reduction of the mental to, or the derivation of the mental from, the neurobiological level of organization is neither descriptively nor explanatorily feasible.

At this juncture, one may insist that the difficulty in decomposing the higher-level mechanisms for cognitive structures, capacities, and processes of language into parts and components that can reach the neurobiological level of organization arises primarily from the conflation of levels of detail with scales of organization inherent in the accounts of Craver (2007) and Bechtel (2008). In fact, Eronen (2015) has accused Craver and Bechtel of this conflation fallacy, pointing out that in many cases entities and activities from two different levels may end up being located at the same level in contributing to a neurobiological mechanism, or conversely, entities and activities from the same level may end up as components at different levels. For example, Eronen provides the example of rod cells which cause bipolar cells to fire and trigger the release of glutamate ions as well. The difference between bipolar cells and the glutamate ions, Eronen argues, cannot be captured in terms of level composition simply because glutamate ions are not components of the bipolar cells and ultimately are at a lower level than the bipolar cells. Here is a case where entities from two levels that are considered distinct interact and form parts of the same mechanism, even though the account of level composition has it that component parts constituting a neurobiological mechanism are at the same level if these parts are not components of one another. Beyond that, that this is not simply a trivial problem is shown by the demonstration of certain logical holes in a proposal that relies on level composition for the identification of neurobiological mechanisms. For instance, two entities from the same level, say x and y, participating in a neurobiological mechanism M may turn out not to be at the same level if *only* direct components, say X and Y, of the mechanism M are considered to constitute M and if x and y are sub-components of X and Y respectively. Given that any two components in the mechanism that are *not* in a component-mechanism relation with each other are supposed to be at the same level, x and Y (or for that matter, y and X) will turn out to be at the same level, contrary to facts. This will obtain because x and Y are not components of each other in virtue of the fact that x is a sub-component of X and hence at a lower level than Y is, and similarly, y and X are not components of each other. An argument that rules out this possibility is spurious because it may happen that some components along with their sub-components in neurobiological mechanisms together participate in the mechanisms concerned. On the basis of these grounds, Eronen believes that scales which are continuous portions of ontological organization determined by size are better suited to partitioning the fabric of nature.

It is important to recognize that the current argument that the higher-level mechanisms for cognitive structures, capacities, and processes of language cannot be decomposed into parts and components that can reach the neurobiological level of organization does not cash in on any restricted or delimited notion of levels of analysis. The term 'level'—wherever it has been used—has been used in its broader and neutral sense which may be taken to mean divisions within the ontological organization determined by size, quality, relevant aspects of phenomena, temporal boundary, or even phenomenology. In this sense, levels can encompass scales as well. So even if one believes that the accounts of Craver and Bechtel have a flawed notion of levels, the present argument remains unaffected since it crucially turns on the quintessential facets of mechanistic-functional explanations in neuroscience.

What quite straightforwardly emerges from the discussion above is the second part of the argument which rests on the point that no compositional relations hold between the mechanisms supporting cognitive structures or processes of language and the parts or components that may be supposed to permit those mechanisms to realize the relevant cognitive structures and capacities by blending into the neural structures at the bottom. As has been pointed out above, from the activities of individual neurons and/or neuronal assemblies, one cannot infer that it is the functional brain areas that are actually interacting with one another to construct a mechanism, even if the brain regions specific to certain brain functions are known to perform certain cognitive functions. For (neuro)biological entities and/or activities can be expected to participate in and hence constitute (neuro)biological mechanisms, and if so, it is not clear how (neuro)biological entities and/or activities can *in themselves* constitute mechanisms for cognitive structures and processes of language. This blocks the neurobiological level from moving up in scale to hold constitution relations with the cognitive.

Hence constitution relations fail to obtain in both upward and downward directions across the neural and cognitive scales.

The picture projected may have one believe that cognitive structures, processes, and associated capacities reflected in language are then 'ethereal' entities and activities, inasmuch as they are neither derived from nor instantiated in the neurobiological base. It must be stressed that the present book does not advocate the view that cognitive structures, capacities, and processes reflected in language are outside the realm of brain functions. One need not jump to the conclusions in this regard. In fact, the presence of the word 'cognitive' here indicates that something pertaining to the brain is associated with the structures, capacities, and processes reflected in language. We may at least judge it appropriate to point out that understanding the representational properties of linguistic structures within the constraints circumscribed by the linguistic capacity has profound implications for an understanding of the brain basis of language. That is because the representational properties of linguistic structures emerge only when language comes embedded in the language faculty which is supposed to be anchored in the neurobiological substrate. This certainly squares with the way a good deal of neurolinguistic research is usually conducted. The other side of this approach involves moving from an understanding of how the brain functions to see how the representational properties of linguistic structures can be accommodated within the neurobiological infrastructure. Both these facets of the brain-mind interfacing are validly accepted by researchers in neurosciences and cognitive science. In a nutshell, the present book does not dispute that a neurobiological basis of the cognitive capacity for language exists, since the language capacity cannot get off the ground without being rooted in the neurological machinery. As a matter of fact, evolutionary mechanisms may have worked to create a neurological space for the emergence of the language capacity by having the foundational machinery of the language capacity wired into the genome (Bickerton 2014b).

What this book, however, disagrees with is that the representational properties of linguistic structures and cognitive structures and mechanisms of language can be derived in a deductive manner from, or instantiated like the software in, the neurobiological substrate. In this respect, the view to be bolstered in this work differs from the *functionalist* approach in (theoretical) cognitive science that adopts the stance that the mind is like the software that is to be studied by abstracting away from the details pertaining to the stuff in which the software is implemented (see Block 1995; Fodor and Pylyshyn 2015). The reason this is so is that the idea this book will press is *not* that no matter how much we ultimately understand about the brain, this will not tell us how the neurobiological can inform what the cognitive can be. A deeper understanding of the brain may prove useful, especially when the direction of explanatory fit is from the mental to the neurobiological. More importantly, the present view does not lend itself to being compatible with the argument from *multiple realizability* (which can be stated as the idea that mental states and functions can be realized in various kinds of substance both biological and non-biological, and hence describing them at the computational/functional level by abstracting away from multiple kinds of substance is fully legitimate). The argument from multiple realizability forms the backbone of functionalism which the present view does not make much of, precisely because describing mental states, structures, and functions (including those that are reflected within and through language) at the computational level is too trivial to have any explanatory import (see Polger and Shapiro 2016; Mondal 2017). After all, a computational-level description of mental states, structures and functions for language is as good as a computational-level description for memory or vision or even emotion. Any cognitive ability or process can be conceived of as a computation detached from the underlying mechanisms of implementation. This is exactly what undercuts the uniqueness of a given cognitive capacity because a cognitive capacity ends up being as *functional* or computational as any other. But this cannot be upheld in a justifiable sense, in that the capacity for visual or olfactory cognition, for example, may be more attached to the underlying mechanisms of implementation than the capacity for, say, social cognition or even language. Therefore, it needs to be observed that the cognitive structures, capacities, and processes of language emerge *only when* the linguistic system as a whole develops as part of the language faculty anchored in the neurobiological substrate. Hence a blind acceptance of substance-independence buttressed by multiple realizability is not what the present book aims to push for. Rather, this book will strongly argue that even though the cognitive structures, capacities, and processes of language emerge only when the linguistic system as a whole develops as part of the language faculty, many properties of the linguistic system take a form of their own when language is acquired, used, and transmitted to other members within and across communities. Therefore, the properties of the linguistic system are intrinsic to the system as a whole, although the activating conditions for these properties to emerge are mediated rather than constituted by the neurobiological organization.

This book will attempt to impugn the emerging doctrine of the hardware realism or simply 'biologism,' which consists in assigning a special epistemological and/or ontological status to the biological substance. While functionalism moves away from the hardware, biologism approximates or gets closer to the hardware level of description, that is, to the neurobiological level of description. Functionalism on the one hand trivializes the hardware level of description on the lower scale, and on the other hand magnifies the higher level of description by way of the projection of cognitive functions couched in terms of computations ultimately realized in the hardware. If functionalism makes much of the generality of cognitive processes, the hardware realism inherent in biologism makes much of the specificity of the biological substance. However, both functionalism and biologism are in accord in at least one respect. Both functionalism and biologism make little of the unique sufficiency of the logical texture of any cognitive sub-system. To put it in other words, both functionalism and biologism in their attempts to hold on to the primacy of a single level of description which is general enough end up neglecting the unique principles and logical structures of natural language. This leads to a general ignorance about the epistemological potential of natural language in revealing the fabric of cognition. Beliefs about the efficacy of computations in unearthing the ultimate truth about the axiomatic system of natural language on the one hand and expectations that the neurobiological substance can fill up the explanatory gaps in understanding the logical texture of language on the other often go together. Although the central thesis of this book supports neither functionalism nor biologism, the critique to be developed in the succeeding chapters of this book will not simply rest on an attack on functionalism, to be clearer. It will go on to debunk the fundamental assumptions advocated by those who think many answers about why language is the way it is can be found in biology.

At this point, a caveat needs to be stated. The thesis of the book is not that one needs to bridge the gap between linguistic cognition and the underlying biological substrate. Rather, it is palpably clear that bridging the gap between linguistic cognition and the underlying biological substrate is the central concern of the current trend in biolinguistics and also in cognitive science research on the relation between language and cognition. This forms the set of background assumptions that the proposed book only presents for the preparation of the critique. As a matter of fact, the present book aims to cast doubt on this research program. What is distinctive about this research program can be clearly discerned in the following passage quoted from Berwick and Chomsky (2011).

> From the biolinguistic perspective, we can think of language as, in essence, an "organ of the body", more or less on a par with the visual or digestive or immune systems. Like others, it is a subcomponent of a complex organism that has sufficient internal integrity so that it makes sense to study it in abstraction from its complex interactions with other systems in the life of the organism. *In this case it is a cognitive organ, like the systems of planning, interpretation, reflection, and whatever else falls among those aspects of the world loosely "termed mental", which reduce somehow to "the organical structure of the brain"*, in the words of the eighteenth-century scientist and philosopher Joseph Priestley. (emphasis added, p. 20)

More particularly, the present book will argue for a view of cognition that can be informed independently and more adequately by the study of the linguistic system per se, especially when the goal is to understand the nature of cognition via language. But the way this can be accomplished is not by *biologizing* the mental, as is the case with Chomsky (2003), who believes that the mental cannot be reckoned to be non-biological since the mental constitute an aspect of the neurobiological with there being no independent and coherent notion of the (neurobiological) substance (or simply, the brain) segregated from the mental. This becomes a pressing concern, especially in view of recent biolinguistic forays into the nature and form of language seek to boldly map the basic texture of human language, especially of the human language faculty (in terms of its linguistic structure and operations) onto patterns of biological structures (facets of neural wiring efficiency, loci of genetic variation for the neural circuits subserving language, etc.) (see Jenkins 2001; Di Sciullo and Boeckx 2011). Although this program is rooted in the linguistic tradition that has celebrated the autonomy of abstraction of linguistic structures, the deeply entrenched riddles and unsettling problems of what Chomsky (2000) often refers to as 'unification' of facets of the cognitive and the biological need to be (re)evaluated in the light of the concerns raised here. This interestingly relates to what Lamb (1999) calls the 'transparency illusion' which makes us tend to think that what appears outside a complex system as an output of processing in the system must also reside inside that system either as a whole or in parts. If the mental is no different (at least in an indisputably substantive sense) from what is (neuro)biological, as per the dictates of biologism, then what is distinctively mental invariably falls inside biological categories. This is exactly what the transparency illusion unmasks. We shall thus uncover many of the fallacies as consequences springing from this illusion.

The overall idea to be expanded on in this book appears to be in accord with what Poeppel and Embick (2005) have argued for when pointing to two fundamental problems in relating the linguistic categories to the neurobiological ones. One is the problem of granularity (the different sizes of linguistic and neurobiological structures; for example, the finer distinctions among rules, principles, and constraints versus the coarse-grained analysis by means of imaging studies or ERP techniques), and the other is the problem of ontological incommensurability (the fundamental divide between linguistic structures and representations and neurobiological structures). Needless to say, these two problems still plague any hypothesis put forward that apparently accounts for the way the linguistic can be matched with the neurobiological. In the face of the trouble, one cannot reasonably maintain that the translation of the linguistic categories into the neurobiological categories can be made easier by having a series of cognitive categories sandwiched between the two levels of description. The problem is much more perplexing and paralyzing than Poeppel and Embick seem to have thought,[4] for it is not merely the form of linguistic structures but also the form of cognition that is *constitutively* informed by language which invites the same kind of problems deriving from the granularity mismatch and ontological incommensurability. One plausible way of relating the form of cognition constituted by language to the neurobiological structures is to regard conceptual schemas[5] for linguistic categories as the building blocks of the linguistic structuring of cognition so that these schemas can be mapped onto the spreading activation models of actual neural systems (Feldman 2006). Even though we shall observe, as we move to Chapter 4, that many of the unique properties of natural language can be conceptualized in terms of general schemas, the logical richness of such schemas will nevertheless admit of the problems of the granularity mismatch and ontological incommensurability. That is because conceptual or linguistic schemas that capture a plethora of logically possible generalizations are not necessarily neurobiological entities—nor are the distinctions and constraints in schemas obviously maintained in the same way in the neuronal systems or in their models (probabilistic or otherwise). This is typical of what Hickok and Small (2016) call 'the mind/brain fallacy' which consists in identifying conceptual models with neuronal models.

Significantly, if the postulated relations of derivation or instantiation between the form of cognition that is constitutively informed by language and the neurobiological organization do not exist, there is nothing in principle that prevents the form of cognition made available by language from being assigned a status which licenses looking inside the cognitive space through properties intrinsic to the linguistic system.

---

[4]It should be mentioned that the alternative that Poeppel and Embick (2005) propose is functionalism when they note that linking hypotheses are missing—which the present book disapproves of. This is another way of stating that functionalism/computationalism is not the only alternative to the problem that arises from the absence of linking hypotheses for a matching between the linguistic and the neurobiological.

[5]This approach applies the notion of conceptual schemas for linguistic structures to the Neural Theory of Language (NTL) project by drawing upon insights from Construction Grammar (Goldberg 2006) and fundamental tenets of Cognitive Grammar (Langacker 1987, 1999; Talmy 2000).

Plus it is no use urging that even if the relations of derivation or instantiation of the linguistically organized form of cognition with respect to the neurobiological organization are not currently incorporated in any account available, we may arrive at something in future. This optimism may arise, in particular, from an argument that plays up the logical possibility of this bright scenario despite the prevailing dampening state of affairs within biology. To be clear, the relations of derivation or instantiation of the form of cognition in connection with the neurobiological organization can also bring forth logical inconsistencies which are barely discerned at present. This book will also attempt to show what these inconsistencies could be.

Furthermore, the goal of this work is not also to investigate the nature of brain computations by elevating them to a higher level or scale at which the representational properties of linguistic structures can be located, contrary to what Ballard (2015) seems to think, chiefly because imposing the property of computations on brain functions does in no way grant a special status to linguistic structures vis-à-vis their neurobiological organization. We find computations running all throughout after all—they have just been factored into different scales. That is, an attempt to understand the representational properties of linguistic structures by abstracting away from the neural details by way of crossing a series of levels/scales within the hierarchical organization of the brain contributes nothing whatsoever to how relations of derivation or instantiation may hold within the brain with respect to the representational properties of language structures. As far as the neural organization of language is concerned, both relations of derivation and instantiation are insufficient to account for the representational properties of linguistic structures and cognitive structures and mechanisms of language. In other words, there is more to linguistic structures than can be walled in by the relations of derivation and instantiation with respect to the neural organization of language. The claim to be buttressed in the rest of this book is in fact much stronger than this. The book would also argue that the (representational) properties of the linguistic system as a whole, besides being necessary, more than suffice to hold the right key to our understanding of the fabric of cognition, whereas understanding the relations of derivation and instantiation with respect to the neural organization of language is neither sufficient nor necessary. This entails that nothing over and above the (representational) properties of the linguistic system is required for penetrating into the cognitive machinery, although an understanding of the neural organization of language may be helpful just like an understanding of the brain may prove helpful for what we may understand about the brain by studying the products of brains such as art, music (Zeki 2008).

Even if the goal of the book, one the face of it, may sound pessimistic, in fact it is the other way round. The book urges that we may do better by delving into the linguistic system and studying the properties of linguistic structures intrinsic to this system, instead of looking for relations of derivation or instantiation within the biological substrate with reference to the representational properties of language structures. Thus, this presents a picture of language–cognition relations which is much richer than has been assumed by many linguists and cognitive scientists alike. It happens to be the case that we are yet to understand, let alone exploit, the potential which language as a cognitive system offers. Thus this book will make out a case for harnessing the full potential of natural language in unlocking the secrets of cognition. No doubt this will necessitate overhauling the whole system of thinking oriented around the relationship between biology and cognition. As we make progress toward a deeper understanding of nature than is available now, it appears that many phenomena that have thus far seemed to resist integration into the fabric of nature are now amenable to unification within the natural sciences. Linguistic and cognitive sciences are now on the verge of such theoretical integration in the sense that linguistic and cognitive categories are now beginning to be treated in much the same way as other by-now-accepted biological categories. This is vindicated by the emergence of enterprises such as neurolinguistics, biolinguistics, cognitive biology. Attempts at unification are not in themselves unjustified as they give rise to myriad new challenges that (can) exercise the minds of many scholars. However, challenges may also erupt into insurmountable difficulties and unmask or unpack a number of deep conundrums and fundamental quandaries which are barely recognized as such. Attempts to discover the relations of derivation or instantiation for the form of linguistically formatted cognition within biology have yielded to this state of affairs, as this book will argue.

The goal of the book is not to demonstrate that biological theories are characterized by fundamental flaws, to be sure, but biology seems to lack what the linguistic system itself possesses. Therefore, the goal is to depict the linguistic system as the most adequate bridge that can hook itself into the fabric of cognition, and to that end, this book will draw upon insights from linguistics, neurobiology, cognitive psychology, computer science, and philosophy. Overall, this has crucial consequences not merely for cognitive neuroscientific or genetic studies on language but also for the very nature of cognition. When much of cognitive science and a whole lot of research in empirical biolinguistics are gradually moving toward the underlying biological base for descriptive and explanatory advantages, the present book aims to call into question this trend by showing that there are no compelling logical grounds for supporting this move. Rather, the examination of a range of linguistic phenomena such as variable binding, quantification, complex predicates, word order projects a picture that is antithetical to the current tendencies in biological studies on language and cognition. Hence this book proposes to integrate a rich understanding of the linguistic organization of, rather than mere contribution to, cognition into a new framework of language–cognition relations. The arguments to be consolidated will become clearer as we proceed to spell out how the linguistic system can qualify as the right linking part for the constitution of the cognitive superstructure. This will be done by penetrating into the logical organization of natural language in order to show how this can uncover hidden cognitive structures as they are constitutive of linguistic structures. The presentation will not attempt an exhaustive or comprehensive analysis of such linguistic phenomena in strictly linguistic terms; rather, they will be presented at a depth that allows for appropriate cognitive consequences to be drawn. The arguments have been pitched at a fairly general level for audiences coming from diverse backgrounds in cognitive sciences. In a nutshell, the book is intended for anybody who bothers about the way language connects to the fundamental structures inside our innermost realm. The book has been made as accessible as has been possible.

The book is organized as follows. The next chapter will examine the ways in which linguistic cognition can be supported by the biological scaffolding and thereby show where the problems lie. Chapter 3 will probe into the question of whether language derives from cognition or cognition itself derives from language. This exercise is believed to smooth the way for a reasonable understanding of cognitive structures having a reflex in linguistic structures. Chapter 4 will thus show how we can gain entry into the realm of cognitive structures when linguistic structures are viewed as requiring nothing other than what linguistic structures in themselves constitute. Chapter 5 will piece together the consequences that emanate from the new view expressed and developed in the preceding chapters and then offer apposite reflections on the nature of language–cognition relations. Hopefully, this will help trace the roots of the mind to something that may ultimately unravel the puzzles revolving around the actual nature of language-cognition relations.

References

Ascoli, G. A. (2015). *Trees of the Brain, Roots of the Mind.* Cambridge: MIT Press.
Ballard, D. H. (2015). *Brain Computation as Hierarchical Abstraction.* Cambridge: MIT Press.
Bechtel, W. (2008). *Mental Mechanisms: Philosophical Perspectives on Cognitive Neuroscience.* London: Routledge.
Berwick, R., & Chomsky, N. (2011). The biolinguistic program: The current state of its development. In A. M. Di Sciullo & C. Boeckx (Eds.), *The Biolinguistic Enterprise: New Perspectives on the Evolution and Nature of the Human Language Faculty* (pp. 19–41). New York: Oxford University Press.
Bickerton, D. (2014a). Some problems for biolinguistics. *Biolinguistics, 8,* 73–96.
Bickerton, D. (2014b). *More than Nature Needs: Language, Mind and Evolution.* Cambridge, MA: Harvard University Press.
Bickle, J. (1998). *Psychoneural Reduction: The New Wave.* Cambridge: MIT Press.
Bickle, J. (2003). *Philosophy and Neuroscience: A Ruthlessly Reductive Approach.* Berlin: Springer.

Block, N. (1995). The mind as the software of the brain. In E. E. Smith & D. N. Osherson (Eds.), *An Invitation to Cognitive Science: Thinking* (Vol. 3, pp. 377–425). Cambridge: MIT Press.

Bouchard, T. J. (2007). Genes and human psychological traits. In P. Carruthers, S. Laurence, & S. Stich (Eds.), *The Innate Mind: Foundations and the Future* (Vol. 3, pp. 69–89). New York: Oxford University Press.

Brown, C. M., & Hagoort, P. (2000). The cognitive neuroscience of language: Challenges and future directions. In C. M. Brown & P. Hagoort (Eds.), *The Neurocognition of Language* (pp. 1–14). New York: Oxford University Press.

Chomsky, N. (2000). *New Horizons in the Study of Language and Mind*. Cambridge: Cambridge University Press.

Chomsky, N. (2003). Reply to Lycan. In L. M. Antony & N. Hornstein (Eds.), *Chomsky and His Critics* (pp. 255–263). Oxford: Blackwell.

Churchland, P. S. (1986). *Neurophilosophy: Toward a Unified Science of the Mind/Brain*. Cambridge: MIT Press.

Churchland, P. S., & Sejnowski, T. J. (1992). *The Computational Brain*. Cambridge: MIT Press.

Clifton, C. (2000). Evaluating models of human sentence processing. In M. W. Crocker, M. Pickering, & C. Clifton (Eds.), *Architectures and Mechanisms for Language Processing* (pp. 31–55). Cambridge: Cambridge University Press.

Craver, C. F. (2007). *Explaining the Brain: Mechanisms and the Mosaic Unity of Neuroscience*. New York: Oxford University Press.

Di Sciullo, A. M., & Boeckx, C. (Eds.). (2011). *The Biolinguistic Enterprise: New Perspectives on the Evolution and Nature of the Human Language Faculty*. New York: Oxford University Press.

Egan, F. (2017). Function-theoretic explanation and the search for neural mechanisms. In D. M. Caplan (Ed.), *Explanation and Integration in Mind and Brain Science* (pp. 145–163). New York: Oxford University Press.

Eronen, M. I. (2015). Levels of organization: A deflationary account. *Biology and Philosophy, 30*, 39–58.

Feldman, J. A. (2006). *From Molecule to Metaphor: A Neural Theory of Language*. Cambridge: MIT Press.

Fodor, J., & Pylyshyn, Z. W. (2015). *Minds Without Meanings*. Cambridge: MIT Press.

Frazier, L., & Clifton, C. (1996). *Construal*. Cambridge: MIT Press.

Garnham, A. (2005). Language comprehension. In K. Lamberts & R. L. Goldstone (Eds.), *Handbook of Cognition* (pp. 1–5). London: Sage.

Gazzaniga, M. (2018). *The Consciousness Instinct: Unraveling the Mystery of How the Brain Makes the Mind*. New York: Farrar, Straus and Giroux.
Godfrey-Smith, P. (2007). Innateness and genetic information. In P. Carruthers, S. Laurence, & S. Stich (Eds.), *The Innate Mind: Foundations and the Future* (Vol. 3, pp. 55–68). New York: Oxford University Press.
Goldberg, A. (2006). *Constructions at Work: The Nature of Generalization in Language*. New York: Oxford University Press.
Goldberg, A. (2019). *Explain Me This: Creativity, Competition, and the Partial Productivity of Constructions*. Princeton: Princeton University Press.
Graham, S. A., & Fisher, S. E. (2015). Understanding language from a genomic perspective. *Annual Reviews of Genetics, 49*, 131–160.
Hickok, G., & Small, S. A. (2016). The neurobiology of language. In G. Hickok & S. A. Small (Eds.), *Neurobiology of Language* (pp. 3–12). Amsterdam: Elsevier.
Ingram, J. C. (2007). *Neurolinguistics: An Introduction to Spoken Language Processing and Its Disorders*. Cambridge: Cambridge University Press.
Jenkins, L. (2001). *Biolinguistics: Exploring the Biology of Language*. Cambridge: Cambridge University Press.
Johnson, G. (2009). Mechanisms and functional brain areas. *Minds and Machines, 19*, 255–271.
Katz, J., & Postal, P. (1991). Realism vs. conceptualism in linguistics. *Linguistics and Philosophy, 14*(5), 515–554.
Lakoff, G. (1987). *Women, Fire and Dangerous Things*. Chicago: Chicago University Press.
Lamb, S. M. (1999). *Pathways of the Brain: The Neurocognitive Basis of Language*. Amsterdam: John Benjamins.
Langacker, R. (1987). *Foundations of Cognitive Grammar*. Stanford: Stanford University Press.
Langacker, R. (1999). *Grammar and Conceptualization*. Berlin: Mouton de Gruyter.
Levine, R. (2018). Biolinguistics: Some foundational problems. In C. Behme & M. Neef (Eds.), *Essays on Linguistic Realism* (pp. 21–60). Amsterdam: John Benjamins.
Luria, S. E. (1973). *Life, the Unfinished Experiment*. New York: Charles Scribner's Sons.
Lycan, W. G. (1987). *Consciousness*. Cambridge: MIT Press.
Lyon, P. (2006). The biogenic approach to cognition. *Cognitive Processing, 7*(1), 11–29.

Mandler, G. (2002). Origins of the cognitive (r)evolution. *Journal of the History of the Behavioral Sciences, 38,* 339–353.

Mitchell, D. C., Brysbaert, M., Grondelaers, S., & Swanepoel, P. (2000). Modifier attachment in Dutch: Testing aspects of Construal Theory. In A. Kennedy, R. Radach, D. Heller, & J. Pynte (Eds.), *Reading as a Perceptual Process* (pp. 493–516). Oxford: Elsevier.

Mondal, P. (2012). Can internalism and externalism be reconciled in a biological epistemology of language? *Biosemiotics, 5,* 61–82.

Mondal, P. (2014). *Language, Mind, and Computation.* London: Palgrave Macmillan.

Mondal, P. (2017). *Natural Language and Possible Minds.* Amsterdam: Brill.

Nagel, T. (2012). *Mind and Cosmos: Why the Materialist Neo-Darwinian Conception Is Almost Certainly False.* New York: Oxford University Press.

Northoff, G. (2018). *The Spontaneous Brain: From the Mind-Body to the World-Brain Problem.* Cambridge: MIT Press.

Pinker, S. (1997). *How the Mind Works.* New York: W. W. Norton.

Poeppel, D., & Embick, D. (2005). Defining the relation between linguistics and neuroscience. In A. Cutler (Ed.), *Twenty-First Century Psycholinguistics: Four Cornerstones* (pp. 103–118). Mahwah, NJ: Lawrence Erlbaum.

Polger, T. W., & Shapiro, L. A. (2016). *The Multiple Realization Book.* New York: Oxford University Press.

Postal, P. (2003). Remarks on the foundations of linguistics. *The Philosophical Forum, 34,* 233–251.

Pulvermüller, F. (2018). Neural reuse of action perception circuits for language, concepts and communication. *Progress in Neurobiology, 160,* 1–44.

Sag, I., Boas, H. C., & Kay, P. (2012). Introducing sign-based construction grammar. In H. C. Boas & I. Sag (Eds.), *Sign-Based Construction Grammar* (pp. 1–29). Stanford: CSLI Publications.

Sanz, M. (1996). *Telicity, Objects and the Mapping onto Predicate Types: A Cross-Linguistic Study of the Role of Syntax in Processing.* Doctoral dissertation, University of Rochester, New York.

Sanz, M. (2013). The path from certain events to linguistic uncertainties. In M. Sanz, I. Laka, & M. K. Tanenhaus (Eds.), *Language Down the Garden Path: The Cognitive and Biological Basis for Linguistic Structures* (pp. 253–262). New York: Oxford University Press.

Silva, A. J., Bickle, J., & Landreth, A. (2013). *Engineering the Next Revolution in Neuroscience: The New Science of Experiment Planning.* New York: Oxford University Press.

Stich, S. (Ed.). (2007). *The Innate Mind: Foundations and the Future* (Vol. 3, pp. 69–89). New York: Oxford University Press.

Stoljar, D. (2010). *Physicalism*. New York: Routledge.

Talmy, L. (2000). *Towards a Cognitive Semantics*. Cambridge: MIT Press.

Tommasi, L., Nadel, L., & Peterson, M. A. (2009). Cognitive biology: The new cognitive sciences. In L. Tommasi, M. A. Peterson, & L. Nadel (Eds.), *Cognitive Biology: Evolutionary and Developmental Perspectives on Mind, Brain and Behavior* (pp. 1–13). Cambridge: MIT Press.

Townsend, D. J., & Bever, T. G. (2001). *Sentence Comprehension: The Integration of Habits and Rules*. Cambridge: MIT Press.

Waddington, C. H. (1940). *Organizers and Genes*. Cambridge: Cambridge University Press.

Zeki, S. (2008). *Splendors and Miseries of the Brain: Love, Creativity, and the Quest for Human Happiness*. Oxford: Blackwell.

2
Biological Foundations of Linguistic Cognition

As we undertake to understand the relationship between the neurobiological substrate and linguistic structures, we wonder how the level of knowledge structures of language and psycholinguistic procedures on the one hand and the neurobiological level on the other can be related. The first chapter of the book has argued that this relation—whatever it may eventually turn out to be—cannot be couched in terms of relations of derivation and instantiation. Ascertaining how this argument may work in actual terms requires that we first figure out where the biological roots of linguistic cognition (may) lie. Hence, it is necessary to examine the biological foundations of language and cognition. Modern thinking on the biological foundations of language and cognition can be traced to Lenneberg (1967), who aimed to set on a firmer ground the connection between linguistic structures and the acquisition of language, and their roots in various biological structures, mechanisms, and processes. The general idea has been to chart the paths that can link biological structures, mechanisms, and processes to distinct components of the language capacity in which language becomes manifest through language development, language use, and neurological disruptions. Lenneberg was careful to point out that cognitive structures, representations, and mechanisms that utilize linguistic structures and representations cannot be directly mapped onto the relevant biological structures, mechanisms, and processes. For example, the structure of categorization as a cognitive process linguistically formatted and (re)structured through naming in language (by means of noun phrases, for instance) is not presented as directly instantiated in any biological structures, mechanisms, and processes.

While it is tempting to treat the domain of knowledge structures for language along with the associated psycholinguistic procedures as a biologically instantiated mental organ, it is possible that the level of knowledge structures for language and the associated psycholinguistic procedures is quite distinct from the level or scale at which the neurobiological substrate is to be located. Now the way these two levels could be very distinct may also rest on the measure of the granularity in terms of which the neurobiological description is couched. It is abundantly clear that the grain 'size' must not be raised to the scale of gross neuroanatomical regions since a simple mapping from linguistic structures and representations to certain brain regions does not furnish a mechanistic explanation, to say the least (Marshall 1980; also see Grimaldi 2012). At worst it provides no explanation. Although it seems evident that we must go down the scale that encompasses the neurobiological substance, it is not clear how to draw up adequately framed relations between a function that can link to linguistic structures or representations, and the underlying neurobiological structures. One way of approaching this issue is to take recourse to the computational/functional level description of linguistic structures, representations, and operations that use such representations. The talk of a computational/functional level distinct from the neurobiological level may be conceptualized in terms of Marr's (1982) three-level schema wherein the top level provides the functional specification dominating the algorithmic level specifying mechanisms or algorithms responsible for the encoding of functions, and the algorithmic level in turn dominates the level of implementation responsible for the instantiation of mechanisms. As a matter fact, Embick and Poeppel (2014) think it is plausible that certain structures at the implementation level, that is, at the level of neurobiological structures, would offer explanations for the presence of specific classes or types of linguistic structures, representations, and operations. They consider it necessary to *integrate* the neurobiological level with the level of linguistic structures and representations in an explanatory fashion, crucially because they regard computations as the bridging link that can put the neurobiological structures in touch with the linguistic representations and operations. Thus, for them, particular brain structures may possess Turing-machine type of computational power (which can be cashed out in terms of operations of reading, writing, and modifying symbols in the brain) but are actually specialized for the 'computation' of specific classes of linguistic representations and structures. This view of brain operations appears to be bolstered by the view adopted in Gallistel and King (2009) who have championed the idea that the brain much like a Turing machine must have a kind of 'read-write' memory for the execution of many cognitive processes. From this perspective, it appears that neural computations can sort of 'scale up' to the level of linguistic structures and representations because operations on such structures and representations are also computations. So it becomes easier to understand that the implementation-level (or neurobiological) computations can be a good fit for certain classes of linguistic computations that we locate at the computational/functional level, for after all both types of computations fall under the same formal category of operations that can be defined by Turing machine operations.

But this is ultimately misguided for two reasons. First, even though we define certain computable functions that can operate on a class of linguistic structures and representations, such computable functions may have nothing to do with the linguistic structures and representations concerned. That is because such linguistic structures and representations may have irreducible linguistic properties that render the computational characterization simply redundant and spurious (see Mondal 2014). For instance, linguistic constructions involving the English conjunction 'and' can be described by defining a computable operation akin to the operation of addition on a class of phrases to which 'and' applies. So we may write *and* (NP^+, NP) = NP— which states that the conjunction 'and' is a *function* that takes exactly two noun phrases (NP) or combines more than one NP with a single NP and then returns another noun phrase. Here '+' attached to the first NP is a regular expression meaning one or more NPs. This covers many English expressions such as 'John, Mary and the neighbors,' 'Sony, Samsung, Amazon and Apple,' 'the bikes and the bats,' 'the bikes and the bats and the pads,' 'the bikes and the bats and the pads and the scooters,' etc. The fundamental problem with this is that this does not account for many linguistically valid overlapping groupings in conjoined NPs. For example, 'the bikes and the bats' in 'the bikes and the bats and the pads' may form a separate group which overlaps with another group formed by 'the bats and the pads.' The meanings of both these groups can coexist in the whole conjoined NP 'the bikes and the bats and the pads.' This possibility cannot be straightforwardly derived from a description of the linguistic rule involving 'and' as a computable function (see also Levelt 2008). This generalizes to scores of other linguistic phenomena that are expected to come under the definition of computable functions. Foisting computations on linguistic structures and rules serves no significant linguistic purpose, and hence, this computational characterization is entirely vacuous. Second, even if it is supposed that neural circuits execute computations, it is doubtful that such computations are constrained by classes of linguistic representations and rules when such representations and rules are defined on the axiomatic system of language and not on the mental processing of language in real time. That is to say that the computations that may be supposed to be carried out by neural circuits may have nothing whatever to do with linguistic representations and rules attributed to the axiomatic system of language just like such computations may have nothing to do with the mathematical/logical properties of a system of algebra. What is *at most* plausible is that such neural computations can constrain and also be partly constrained by the psycholinguistic mechanisms employed in language processing, especially when linguistic structures are built and comprehended piece by piece by utilizing capacities of memory, vision, and motor systems. But, even there it is absurd to think that assumptions about computational operations can be straightforwardly extracted or carried over from the demarcation of specific neurobiological structures. In other words, it is not easy to determine whether neurons, neural assemblies, networks of neural assemblies, and broad neural regions *all* execute computations. Since it is not easy to fix the sources of computing in the brain, it is not equally easy to ascertain from which level of implementation (in terms of grain size) computations may arise and 'scale up' to the level of psycholinguistic procedures utilized in language processing. If computations are restricted to only neurons and neural assemblies, an explanation of how such computations move up to the higher level of psycholinguistic mechanisms and give rise to the unique character of such mechanisms is warranted. But, even if such an explanation is provided, it is unlikely that the connections established between the neurobiological level and the level of psycholinguistic mechanisms will be tight enough because any computations restricted to only neurons and neural assemblies have to be transparently mapped on to, or simply pass through, many sizes of neurobiological structures up to the scale of broader brain regions. This is required since psycholinguistic mechanisms cannot be said to operate at the level of neurons or neural assemblies. Clearly, this will ultimately render computations vacuous once computations are 'filtered through' many sizes of brain structures—when computations reach broader brain regions in order to be in touch with psycholinguistic mechanisms, they cease to have their mechanical character simply because brain regions do not carry out any neural mechanisms, as also emphasized in Chapter 1. Therefore, the putative computational link wedged in the gap between neurobiological structures and psycholinguistic mechanisms looks at best suspect.

Further, most people think that some functional/computational level structures can serve as the right elements mediating the link, and these structures are reckoned to be *nothing other than* neural structures described in functional/computational terms (see also Caplan 1984; Brown and Hagoort 2000; Feldman 2006; Hagoort 2014; Sprouse and Hornstein 2016). This is so because these structures are thought to be affected by genetic transmission, environmental and cultural molding, and pathological insults and injuries. However, it is not at all comforting to think that brain functions that can be restructured or disrupted over the period of one's life are hooked to particular representations of linguistic structures, given that in many cases there may exist many-to-many mappings between brain functions associated with linguistic structures and the particular loci of neurobiological structures (Kaan 2009; Hagoort 2009). Thus, it is not unreasonable to suppose that the matching of cognitive structures bootstrapped from linguistic structures with the neural structures cannot be telescoped into a computationally viable unique neural encoding format. That is, many-to-many mappings between brain functions associated with linguistic structures and various neurobiological structures render the projection of functional/computational level structures from the neural structures cumbersome and nugatory, given that the projected functional/computational level structures for linguistic cognition are not uniform with respect to their grounding on the one hand and the grounding structures are not uniformly responsive to the projected functional/computational level structures for linguistic cognition on the other. If the projection of functional/computational level structures from the neural structures is mainly for mediating the link between the cognitive structures and/mechanisms and the neurobiological structures both encompassed by the neurobiological scale overall, it is hard to see how the functional/computational level structures can serve to mediate the link between the cognitive structures and/mechanisms and the neurobiological structures. This is precisely because the postulated projection of functional/computational level structures from the neural structures does *no more than* isolate a facet of the neurobiological scale for the localization of functional roles of cognitive structures and/or mechanisms. This is another way of saying that the many-to-many mappings between cognitive structures bootstrapped from linguistic structures and the grounding neural structures can be described without positing any intermediate-level structures because doing so serves no purpose in revealing the unique properties of cognitive structures and mechanisms that are reflected in and through language.

It is conceivable that specific cognitive structures and mechanisms that are reflected in language are *in themselves* variably manifest in the neural substrate due to variations in genotype–phenotype correlations such that specific genomic targets of specific cognitive structures and mechanisms mediated by language are expressed in neural structures in a cross-modular fashion and specific cognitive functional targets of neurobiological structures are matched by multiple similar genomic expressions across brain regions (see Fisher and Vernes 2015). This possibility is indicative of the presupposition that language–cognition relations must ultimately be rooted in the genetic infrastructure. That this idea is problematic can be revealed by careful scrutiny. The way genomic targets of specific cognitive structures and mechanisms as manifested or reflected in language may be expressed in the brain is quite different from the way specific cognitive functional targets of neurobiological structures can appear in gene expression within the brain. It may be observed that the former lies within the domain of biology, while the latter may not stay within the ontological limits of biology, for the latter, but not the former, is determined brain-externally by humans as intentional agents. Specific cognitive functional targets of neurobiological structures cannot be said to *naturally* fit regularities of genomic expressions and gene regulation networks. The reason for this is that cognitive functions do not pick up neurobiological structures on their own; nor do cognitive structures and/or mechanisms get themselves attached to certain neurobiological structures by means of some unknown mechanisms—genomic or otherwise. Rather, the cognitive functions, structures, and mechanisms traced to a given neurobiological structure or to a number of neurobiological structures are determined by individuating those cognitive functions, structures and mechanisms first and thereby attempting to check how they can be realized in the associated neurobiological structures. So this is a top-down procedure of figuring out how neurobiological structures can target cognitive functions and mechanisms. But the problem that this gives rise to is that this top-down procedure is not something that finds a comfortable place within the laws and constraints of biology. This is not, of course, to endorse the view that biological constraints are irrelevant to this top-down procedure. What this point emphasizes is that this top-down procedure is not *in itself* a biological process or about a biological mechanism, although it may be relevant to biological constraints just as the matching of various other things minds do may be relevant to certain biological facts about the brain.

While there is no reason to make too much of the supposition that specific cognitive functional targets of neurobiological structures are matched by roughly homogenous genomic expressions across brain regions, it appears that we may do better by making explorations into the way genomic targets of specific cognitive structures and mechanisms as assimilated into language are expressed in neurobiological structures. However, this is also illusory for several reasons. First, genomic expressions via genetic regulation (the interaction of regulatory proteins with the DNA to determine how protein-coding genes are expressed) operate at the level of biophysical molecules and cells. No genomic expressions are driven by the coding of specific cognitive structures and mechanisms or directly construct specific cognitive structures and mechanisms. What is at best plausible is that genomic expressions can influence the building of proteins in the parts of the developing organism, thereby controlling and modifying the developmental paths of neurological structures, the trajectories of connectivity growth among neural structures and their layout. This does not in any way motivate the possibility that specific cognitive structures and mechanisms manifest in language can be *directly* targeted by genes or patterns of gene expression. Second, we should also bear in mind that when one talks about causal connections between genetic coding or patterns of gene expression and the manifestation (or the disruption in the manifestation) of cognitive structures and mechanisms, this causal link is to be understood as nothing but a kind of correspondence. These correspondences may also dwindle into mere correlations, if the causal link is couched in probabilistic terms. It must be noted that this issue has nothing substantive to do with whether the relevant correlations between genotypes and cognitive phenotypes as a reflex of language can be measured in terms of additive or non-additive effects of genotypes and environments (see, for discussion, Lewontin 2000; Sesardic 2005). This is because all additive or non-additive effects of genotypes and environments toward the calculation of heritability of certain cognitive traits are based on and thus presuppose statistical or non-statistical conditional norms of causal interactions between genotypes and environments. Even though these causal interactions between genotypes and environments may furnish estimates of heritability calibrated to the evaluation of the *relative* causal contribution of genotypes, in no sense can these causal interactions between genotypes and environments have the genomic mechanisms 'pick up' or select specific cognitive structures and mechanisms that come to be available or manifest in (cognitive) phenotypes.

Besides, even if the causal link specifies a correspondence, the correspondence cannot be understood in mechanistic terms simply because a series of possible biological mechanisms is *imagined* to be wedged between genetic coding with patterns of gene expression and the manifestation of cognitive structures and mechanisms. The absence of a mechanistic explication of this correspondence renders the very notion of correspondence inconsequential and empty of substance. One cannot circumvent the problem by just positing that a black box of intermediate mechanisms exists and mediates the desired correspondence somehow. After all, biological laws and constraints require real biological substance to be functional or operative. This is precisely what is missing in biologically insignificant and substantively vacuous correspondences.

Nevertheless, different accounts of the biological grounding of linguistic cognition exist. To determine whether these accounts fare well in light of the arguments marshaled in this book so far, we need to scrutinize these accounts in detail. With this, we may now proceed to look at these proposals.

2.1 Genetic Foundations of Language and Cognition

We may first look into the ways in which the genetic foundations of language and cognition can be thought to shed light on the nature of linguistic cognition. One of the primary ways of ascertaining how the genetic foundations of language and cognition obtain is to examine the effects of genetic constraints under which the emergence of language and linguistic cognition can be seen to be enveloped. The *critical period hypothesis* is exactly a way of viewing the genetic foundations of language and cognition in this light (Lenneberg 1967). The critical period for language acquisition constitutes a time window for the biological growth or maturation of language and also for the associated cognitive development that either supports it or is supported by language. The time window starts right after the birth of a baby and is supposed to end near the puberty. Since such critical periods exist for the development and maturation of binocular vision or biped walking in humans and the learning of songs in songbirds, it is believed that this underscores the biological grounding of language or linguistic cognition. The most noticeable evidence for the critical period hypothesis comes from the study of a girl named 'Genie,' who had been alienated from all kinds of human contact since the age of 2 by her father and was later discovered when she was 13. Vigorous training in language learning helped her learn the vocabulary of English, but she could never learn the syntactic principles that are valid in English (Curtiss 1994). Although this seems to point to a clear-cut case of a genetic envelope that constrains and limits the time frame within which one can acquire language, this idea has not gone without criticisms.

First of all, the constraining effect of the supposed critical period for language acquisition is not an all-or-nothing affair. In many cases, the critical period in itself seems to offer leeway to embrace the effects of other factors (biological or non-biological) that may counterbalance the time constraints of biological maturation. It has turned out that in language acquisition processes, certain components of language are more than constrained by the time frame of biological maturation, while certain other components are less constrained by such limits as those posed by the critical period. For example, learning semantics and vocabulary is less sensitive to the limits posed by the critical period, whereas learning syntax and even structured domains of sound systems which constitute the formal domain of linguistic rules and principles appears to be under a tighter genetic control (Newport et al. 2001; Friedmann and Rusou 2015). The caveat is that this does not, of course, say much about the biological mechanisms that underlie this relative determination of the learning of various components of language by the critical period. Beyond that, it is not also reasonable to think that the temporal window set by the critical period acts as a switch that is turned off as soon as one reaches a certain age range; rather, it often acts a valve that shows characteristics of sporadic opening and blocking (Herschensohn 2007). Besides, even though the language learning ability, especially for syntax, may be said to gradually decline after adolescence, the very existence of the critical period for language learning ability could be due to various factors having to do with culture (major life transitions in social life during and after adolescence) or interference from the first language or neuronal changes due to learning, or a combination of these factors (Hartshorne et al. 2018).

Second, Sampson (2005) dismisses claims made about the critical period limitations, especially in the case of Genie, on the grounds that Curtiss (1977) in her initial work did not endorse a stronger version of the critical period hypothesis. Plus that Genie ended up not being able to learn the syntactic rules of English was largely, if not wholly, due to several unhappy circumstances that led to her changing foster homes a number of times at regular intervals, as Sampson points out. Additionally, he also contends that there is no reason to believe that the time window for language learning shuts down *completely* after puberty when a good number of people succeed in learning languages even after puberty.

Third, even if one assumes that a critical period for language acquisition exists, this in no way establishes that the properties intrinsic to the linguistic system as well as to linguistic cognition are directly instantiated in the genetic mechanisms that govern what the critical period for language acquisition does. Rather, what this shows is simply that language development, especially the acquisition of syntactic knowledge, moves in sync with the unfolding of relatively tighter genetic constraints. What is ultimately acquired or learnt thus conforms more or less to the contours set by the specific constraints laid out by our genetic endowment. Since all parts of language do not possess a formal character in an equal measure, it is not surprising that matters of word meanings and vocabularies in languages are less sensitive to a tighter genetic control. Plus semantic structures when associated with some general patterns of conceptual systems are also less subservient to genetic constraints. From this, it seems clear that the properties intrinsic to the linguistic system or to linguistic cognition cannot be said to genetically coded or rooted in our genetic foundations, *just because* genetic constraints shape the temporal window within which language can be acquired. An analogy from the domain of binocular vision can illustrate the point that is made here. Even though we are aware that binocular vision develops in humans under a strict genetic control, from this it does not follow that many intrinsic properties of the visual system or visual cognition uncovered by certain kinds of perceptual illusions are directly instantiated in the genetic endowment for our binocular visual capacity.

It needs to be stressed that this issue has nothing to do with the apparently never-ending tension between nativism versus empiricism (see, for recent discussions, Piattelli-Palmarini and Berwick 2012; Evans 2014). When it is stated that the properties intrinsic to the linguistic system or to linguistic cognition do not hold relations of reduction or instantiation with respect to the genetic substrate, the argument does not turn on the question of whether or not we should accept a view that supports a domain-specific genetic endowment for language or a view that seconds the proposal for a domain-general genetic endowment for cognition which provides the scaffolding for language. Notice that all that the present argument underlines is that no relations of reduction or instantiation suffice to have the properties intrinsic to the linguistic system or to linguistic cognition anchored in the genetic substrate. For this, a commitment to the idea that there exists a *minimal* form of genetic contribution to linguistic cognition suffices, although it must, at the same time, be recognized that some advocates of empiricism may also rule out the possibility of there being any genetic contribution to language. From this perspective, the present argument must not be confused with an empiricist argument for a minimal form of genetic control or even no genetic control for language (or language acquisition), for empiricism in itself is not incompatible with the view that the form of linguistic cognition is biologically structured and thus ultimately anchored in available biological mechanisms. That is, the empiricist stance gives more weight to the *non-inherent* source of knowledge structures for domains of cognition including language, and this need not be taken to mean that the interactions with the non-inherent source of structures of knowledge for domains of cognition cannot be biologically governed.

If what has been described above can be taken to be one of the ways of looking into the genetic foundations of language and cognition, there are also other ways in which the genetic foundations of language and cognition can be examined. Twin studies and the study of the genetic etiology of speech and language disorders are such cases. We shall examine both cases one by one. Twin studies are an excellent means of understanding the role of the genetic substrate in mediating cognitive functions or phenotypes, since in twins there is a great deal of overlap in the genetic material. Twins are of two types: monozygotic and dizygotic. Monozygotic twins share 100% of their genetic material, whereas dizygotic twins share almost 50% of their genetic material. The underlying idea is that shared environments of monozygotic twins can contribute to similarities in the resulting cognitive phenotypes in twins, and similarly, unshared environments of monozygotic twins may contribute to differences in the resulting cognitive phenotypes in twins. Notice that this conclusion is harder to make in the case of dizygotic twins, mainly because the co-twins do not share *all* the genetic material and this leaves room for greater non-genetic influences on the resulting phenotypic differences and similarities. At this juncture, it is also crucial to understand that if we find any similarities in the cognitive phenotypes manifested in monozygotic twins, we cannot readily conclude that the relevant similarities are *solely* due to the fully shared genetic material. This is because we have not taken into account the roles of the environments which may be shared or different in kind. Thus, if the environments in monozygotic twins are uncorrelated and hence different in nature, then we may, somewhat safely, infer that the concordance in the cognitive phenotypes in monozygotic twins is due to the fully shared genetic material. This inference unequivocally presupposes that differences in the environments of the monozygotic twins concerned cannot accumulate to form some converging zones of similarities randomly or otherwise that could suffice to provide the explanation for the concordance in the cognitive phenotypes in monozygotic twins.

It has been observed that even if monozygotic twins and dizygotic twins share the same perinatal and postnatal environments, monozygotic twins are found to possess more similar linguistic abilities than dizygotic twins (see Stromswold 2001). Such an observation is made sense of by employing the argument that this becomes possible primarily, *if not exclusively*, because of the fully shared genetic material in monozygotic twins. Now here is an important point to be taken note of. A greater degree of similarity in the linguistic abilities of monozygotic twins than in dizygotic twins can also be attributed, at least in part, to the shared perinatal or postnatal environments within the context of co-twins of the monozygotic type. Stromswold (2006) also shows that genetic factors account for as much as 60% of the differences or similarities in the linguistic abilities of monozygotic twins and argues that perinatal and postnatal environments could also contribute to differences in the cognitive phenotypes that characterize the linguistic abilities in monozygotic twins. It needs to be made clear in this context that perinatal factors include factors that occur right before or after the birth of the child and are within the domain of biological processes. These include intrauterine experiences and also incidents that prove dangerous such as injuries, accidents, malnutrition, and infection, whereas postnatal factors involve sociocultural factors that affect the development of cognitive capacities including language. Stromswold (2006) points out that epigenetic factors (facts about how genes are expressed in organisms) as well as perinatal factors do account for the differences in linguistic abilities in monozygotic twins, whereas postnatal factors may contribute to differences in non-linguistic cognitive abilities in monozygotic twins. Additionally, the effect of shared postnatal environments on the outcomes of the phenotypes of linguistic abilities in monozygotic twins is also found to be minimal in the sense that the heritability estimates of linguistic abilities are essentially identical, regardless of whether monozygotic twins are reared together or apart (see Pedersen et al. 1994). In a nutshell, all this suggests that the emergence of the cognitive phenotypes that characterize linguistic abilities can, to a large extent, be traced to the genetic and other biological factors. Overall, twin studies appear to show that the linguistic capacity and the structure of cognition that is made available by language are ultimately rooted in our genetic or biological substrate.

Be that as it may, this account projects an illusory picture of the relation between linguistic cognition and the underlying genetic substrate. First of all, linguistic abilities in such studies relate to abilities in the individual components of the linguistic system such as syntax, phonology, and vocabulary. It is supposed that any similarities or differences in the scores of the tests conducted with respect to the different components of the linguistic system can be ascribed to the underlying linguistic abilities actually acquired. In other words, it is the underlying linguistic abilities that are correlated with differences or similarities in the test scores for the relevant different components of the linguistic system. On the other hand, various pieces of molecular evidence are gathered to unmask similarities and differences in the genetic substance in twins. Overall, two kinds of inferences are made in the two cases. In the first case, the examination of linguistic abilities in twins is brought to bear upon the genetic similarities (as in the case of monozygotic twins) or dissimilarities (as in the case of dizygotic twins), whereas in the other case the underlying similarities or dissimilarities at the biological level, especially at the genetic level, are carried forward for extrapolations up to the scale of cognitive phenotypes. The first inference carries over a more distorted logic than the second. The reason is that if the examination of linguistic abilities in twins shows that individual co-twins within a zygotic type (the monozygotic type, for example) are similar or dissimilar in their linguistic abilities or a certain pair of twins across the zygotic types is similar or dissimilar in certain linguistic abilities with respect to the other pair, this does not tell us anything consequential about the underlying cause of the respective similarities or dissimilarities. The reason is that the similarities or differences in the linguistic abilities could be due to many co-varying factors inextricably intertwined with one another. If individual co-twins within a zygotic type are similar/dissimilar or a certain pair of twins across the zygotic types is similar/dissimilar with respect to the other pair in certain linguistic abilities, it does not necessarily license the conclusion that the similarities or dissimilarities concerned must be due to the shared or non-shared genetic material.

Aside from that, that there are similarities or differences in the linguistic abilities in twins does not also provide ground for the supposition that the properties of the linguistic system that enables linguistic abilities are instantiated in the genetic material. To link the similarities in linguistic abilities in twins in a non-spurious manner to the genetic similarities, one needs to furnish the pertinent correspondence principles or the bridging constraints that may help connect the two ontologically distinct scales. Clearly, merely pointing to the correspondence between the underlying similarities in the genetic material and the manifest similarities in linguistic abilities in twins is not enough. If one insists that in such a case all that is highlighted is that genetic factors play a role in the manifestation of the relevant similarities or dissimilarities in linguistic abilities. This is as good as saying that genetic factors play a role in the manifestation of *any* phenotype that we identify. Thus, this seems to trivialize the role the genetic substrate is supposed to play in the instantiation of cognitive phenotypes in the genetic infrastructure.

There is also another line of reasoning that is deployed to suggest that if any environmental factors play a role, if any, in contributing to the differences in linguistic abilities in pairs of twins, such environmental factors must be restricted to organism-internal environments and thus are within the domain of biological principles and constraints. So the experimental studies that probe into this matter indicate that twins having certain genetic risks that predispose them to being affected by selective perinatal environmental factors must have shared the same perinatal environmental factors if it is the case that the linguistic development in the twins concerned is affected by certain perinatal environmental factors (see Stromswold 2006, 2008). That is, the greater impact of certain perinatal environmental factors is correlated with a greater discrepancy in the estimates for environmental and genetic factors for language development in pairs of twins. In all, the idea is that if there are any environmental factors over and above the genetic factors that could be said to contribute to differences in linguistic abilities in pairs of twins, these factors must be biological, if not outright genetic. This reasoning is also fallacious. Note that a higher estimate for the shared environmental factors in a certain pair of twins is supposed to be the outcome of the mechanisms through which perinatal environmental factors influence and shape the development of linguistic abilities in twins. These mechanisms are supposed to function the way they would, but the only assumption is that we get the required evidence especially when the relevant pair of twins is accidentally or otherwise exposed to the adverse perinatal shared environment that is sure to affect the twins. It is easy to notice that the explanation involved rests on a post hoc inference about what could have happened to cause the manifestation of a certain linguistic or cognitive phenotype. For, if the relevant estimate for the impact of the shared environmental factors in a certain pair of twins is higher than that for the impact of the genetic factors, from this it does not necessarily follow that the affected linguistic abilities in the twins are, even at least in part, the result of some adverse shared perinatal events *just because* the shared environment estimates for the twins concerned are higher. It is possible, but not at all necessary. After all, the higher shared environment estimates for the twins concerned may also lead one to conclude that the affected linguistic abilities in the twins are due to some factors the twins did not share. For instance, if some perinatal infection is (partially) responsible for the linguistic delays or deficits in a certain pair of twins, then the fact that the shared environment estimates for the twins are higher does not *necessarily* license the inference that the shared event of a perinatal infection having occurred is the cause of the linguistic delays or deficits. This is because it could be the case that the perinatal events that the twins did not jointly share (e.g., distinct toxic effects on the developing fetuses of twins *within the same shared environmental context*) differentially caused the manifestation of the linguistic delays or deficits in the twins, given that linguistic delays or deficits can be (partially) caused by many perinatal factors.

If the case as delineated above holds true, it is no use giving much weight to the argument that all environmental factors that could be said to contribute to differences in linguistic abilities in pairs of twins are perinatal and hence biological. What is at the best plausible is the inference that the affected linguistic abilities in twins are, at least in part, the result of some adverse perinatal events that obtained when the twins were exposed to these events, given that all human babies undergo perinatal experiences of some kind or other. But this does not in any way affirm the inference from a higher shared environment estimate to the necessary contribution to the present linguistic delays or deficits because of some adverse perinatal events that we know obtained in the past. It is like saying that if the estimate of the shared environment for swimming which only two persons, say, A and B, have been learning together is higher than that for some genetic deficits in A and B that may predispose them both to being incapable of making motor movements required for swimming, this fact itself licenses the conclusion that a certain harmful effect of swallowing water during swimming such as respiratory arrest appeared in both A and B, if the respiratory arrest in both A and B is due in large measure to the swallowing of water during swimming. It is easy to recognize that the connection is spurious, precisely because the inference from the greater estimate of the shared environment for swimming for both A and B to the explanation of the cause of the respiratory arrest in both A and B may well be fortuitous.

Therefore, similarities in linguistic abilities in twins cannot be brought to bear upon the genetic similarities (as in the case of monozygotic twins), since such inferences infelicitously cross the boundary between the level of non-biological properties and the biological substrate. Nor can the dissimilarities in linguistic abilities in twins be adequately accounted for by appealing to some biological factors. In each case, it is quite possible that the properties of the linguistic system manifest in the linguistic phenotypes in twins evince aspects of the linguistic system itself which we identify, characterize, categorize, and analyze from outside and which cannot in an appropriate biological sense be determined by or ascribed to either perinatal factors or genetic and epigenetic factors. That is, it may be a category mistake to bring similarities or dissimilarities in linguistic abilities in twins into correspondence with the causal roles and operations of perinatal, genetic, and epigenetic factors. Moreover, the other type of inference from the biological substrate to the form certain cognitive phenotypes assume is also problematic. Thus, when extrapolations from the underlying similarities or dissimilarities at the biological level, especially at the genetic level, in twins are carried up to the scale of cognitive phenotypes, this raises the deeper question of figuring out whether we are mapping the right parts of the genetic substrate (genotypes) onto the right cognitive/linguistic phenotypes. Just because we observe a correlation between a number of molecular deformities or defects in twins and the relevant cognitive/linguistic phenotypes, we cannot always infer that the observed characteristics (defective or otherwise) of the cognitive/linguistic phenotypes are solely due to those molecular deformities or defects in the genome. One cannot a priori rule out the possibility that the observed characteristics (defective or otherwise) of the cognitive/linguistic phenotypes are a by-product of a number of tangled factors of which the ensemble of molecular deformities or defects is just one part. It is equally possible that the observed characteristics (defective or otherwise) of the cognitive/linguistic phenotypes are largely, if not exclusively, caused by certain psychosocial and cultural factors that could have led to the emergence of the observed linguistic phenotypes. It is not just the shared environment that matters the non-shared environment that cannot be just restricted to within-family experiences turns out to be significant even in the case of twins who may undergo different experiences as individual siblings (see Plomin 2011). It is essential to understand that the observed characteristics of any cognitive ability are due to developmental processes that fuse biological and non-biological factors often in stochastic ways which may show 'chaotic' behaviors because certain minor or small changes can ultimately lead to larger effects on the time of attainment of the cognitive ability (see Balaban 2006).

Furthermore, it is not also necessary to count the assessment of linguistic abilities in twins necessarily as a *dispositional* characterization of the form the developing linguistic system in twins takes. When markers of genetic deformity or defects are found at the molecular level and then matched with the observed patterns of linguistic abilities as analyzed in twins, it is not compellingly obvious that the mapping between the observed genetic deformities and the observed characteristics of the linguistic abilities in different linguistic components is smooth enough. No matter how smooth the mapping may look like, the task of appraising linguistic abilities must turn on the *occurrent* states of the linguistic system in which various linguistic abilities are manifested. And if this is the case, it is not clear why the assessment of linguistic abilities, especially when test scores in these abilities are brought into correspondence with genetic deformities or vice versa, must point to the stabilized fixed state of the linguistic system. More importantly, any observed similarities in the linguistic abilities of twins that are attributed to the shared genetic material are compatible with unexamined dissimilarities in the linguistic abilities of twins, in that any situation that reveals similarities in the linguistic abilities of twins as correlated with certain patterns of genetic sharing is also compatible with all the dissimilarities that the observed similarities background. If this is so, it is not also clear why these dissimilarities in the background against which the observed similarities make sense cannot also be said to be correlated in an equal measure with the same patterns of genetic sharing as those that are matched with the similarities. This issue points to a subtle conundrum concerning the mapping between the assessment of linguistic abilities and its translation into terms that can be brought to bear upon estimates of genetic factors.

If the discussion above seems to bring into our focus the severe troubles any account of the genetic foundations of language and cognition is interspersed with, one way of getting around the troubles is to invoke the role of certain genetic disorders in revealing how the genetic infrastructure shapes and is responsible for the appearance of various aspects of linguistic cognition. The underlying idea is that the functional roles of certain biological structures do not always appear to be biologically transparent, especially when the structures concerned function well, but when these structures are disrupted or damaged their functional roles for a given aspect of (linguistic) cognition become more transparent. A number of speech and language disorders such as *specific language impairment*, dyslexia are revealing cases of linguistic disorders most of which have been found to have biological roots. Since the assumption is that most of these disorders can be studied to ascertain how the observed genotype–phenotype relations mediate the manifestation of these disorders, it seems that one may justifiably be inclined to maintain that the genetic causes for the emergence of speech and language disorders provide the natural scaffolding for the biological grounding of linguistic cognition. To the extent that this line of reasoning is cogent, one may also believe that there is no escaping the fact that the genetic substrate underlies all that we observe in the developed (sub)systems of the cognitive machinery.

The best example of this argument comes from the discovery of the gene called FOXP2, which is responsible for a severe form of speech and language disorder that may run through the generations of a single family. Such a case occurred in a family (called the KE family) in London in which half of the family members had a severe form of speech and language disorder (Gopnik 1990; Hurst et al. 1990). When this disorder was discovered, it was soon believed that this disorder must be traced to some underlying genetic basis for language. This genetic basis turned out to be constituted by FOXP2, a gene that was in a mutated form in the KE family and thus assumed to be the underlying 'language gene' given the remarkable patterns of inheritance in the KE family. However, later it turned out that FOXP2 is not simply a language gene; rather, it is a transcription factor that regulates the expression of other genes in brain development, and it is also responsible especially for the control of muscles and articulation (see Vargha-Khadem et al. 1995; Watkins et al. 2002; MacDermot et al. 2005; Kang and Drayna 2011). Plus FOXP2 has been associated with vocalizations in many other primates and other mammals such as rats, birds, and bats (see Enard et al. 2002; Haesler et al. 2004). Insofar as we make a distinction between language and speech (the former being the system that language users access and the latter being the concrete product of the linguistic system in outward behavior), FOXP2 or any gene cannot form the genetic basis of language no matter what FOXP2 is actually responsible for. Further, no conclusion about the genetic instantiation of the linguistic system or of linguistic cognition can be drawn, even if FOXP2 turns out to be crucial for the development of the linguistic system apart from being causally relevant to the construction of the ingredients for speech. Besides, the expression and the role of FOXP2 in many of the language and speech disorders can be shown to be driven by multiple phenotypic foci that are components of certain global developmental disorders which are not consistently found in the individuals/families in which the mutation of FOXP2 has been detected (see Bishop 2006; Newbury and Monaco 2010). This largely undercuts the exclusive functional role of FOXP2 either for language or for speech.

There is also solid reason to believe that the genetic substrate has no way of implementing linguistic cognition, even though most genetic studies conducted to trace out the genetic causes of various observed speech and language disorders may have one believe otherwise. Two types of studies are executed in virtually all genetic studies that probe into the genetic causes of various observed speech and language disorders. These are linkage studies and association studies. Linkage studies identify certain genetic markers which can constitute the genetic identity across a section of individuals within or across families and then map these patterns of genetic identity onto the observed similarities in the phenotypes found in certain disorders. Association studies, on the other hand, isolate certain genetic variants that are supposed or found to play a causal role in the manifestation of the disorders concerned, and these genetic variants can be expected to be found more commonly in those affected than in those not affected by the disorders concerned. Clearly, neither of these two types of studies in genetics furnishes any substantive description of instantiation relations between the genetic substrate and the relevant facets of the linguistic system. Nor does it provide any mechanistic explanation of the observed association or linking that supports the mapping between the genetic substrate and cognitive phenotypes.

The same conclusion holds in a more general sense for any claim that aims to advance the role of the genetic substrate for linguistic cognition. The linguistic disorder that was observed in the KE family is part of a family of disorders commonly called specific language impairment (SLI). Since it is supposed to be a disorder that affects only the linguistic system without any apparent cognitive deficits that may be responsible for the disorder, one may find it tempting to contend that SLI as an exclusively linguistic disorder makes for a strong case for the biological grounding of language. Although SLI in children has often been observed to affect phonology and morphosyntax (Gopnik et al. 1997; van der Lely and Christian 2000), SLI also affects the pragmatic competence in children affected with the disorder—which suggests that the disorder is also associated with the disruption in capacities for inference, reasoning and conventional understanding of linguistic structures in contexts (see Katsos et al. 2011). Additionally, not everyone agrees that SLI is predominantly a language-specific disorder that disrupts the capacity for language, for many of the characteristic features of language deficits in areas of phonology and morphosyntax may be explained in terms of impairments in domain-general cognitive capacities for input processing, attentional focus or the manipulation of resources in memory functions that largely affect the acquisition rather than the use of linguistic structures (Johnston 1997; see also Leonard 2014).

Add to that the problems in relating the components of a complex linguistic phenotype to the genotypic patterns. Beyond that, all too often the components into which a complex linguistic phenotype is fractionalized are not found to be in a unique congruence with certain genotypic patterns of mutations or deformities. Nor do single genes code or are responsible for the isolated expressions of independent facets of the linguistic system in a neat and clearly partitioned fashion. Various facets of a single linguistic phenotype in any such disorder appear to connect to multidimensional points of a lattice within the linguistic system. Thus, for example, even though dyslexia is assumed to be a linguistic disorder that affects abilities in writing and spelling, it is found to involve impairments not only in other linguistic domains such as phonology but also in capabilities that engage in sequencing objects, colors, or shapes. On the basis of these facts, Fisher (2006) concludes that it is illusory to maintain that there are certain 'abstract' genes that encode or are responsible for the direct expression of linguistic cognition, and for that matter, of cognition in general. Hence, it is not surprising to find that any distortion at the genomic scale need not give rise to a distortion in any cognitive or linguistic phenotype or vice versa. This effect is expected only when one thing is not directly instantiated or implemented in another thing. This is not, however, to say that certain genotypic patterns cannot mediate the manifestation of certain abilities or capacities deployed in linguistic cognition, given that we recognize that abilities or capacities deployed in linguistic cognition are distinct from the linguistic system per se or more so from the properties of the linguistic system. This indicates that certain abilities or capacities deployed in linguistic cognition may be easy to relate to the underlying biological mechanisms, whereas the same may not (always) hold true for the linguistic system as a *representational* organization of human cognition.

In a nutshell, if the genetic underpinning for the properties of the linguistic system in particular cannot even lift the case for the biological scaffolding of the linguistic system onto a level which is fully compatible with our understanding of mechanistic principles if biology, one may now point out that the safest place to look for the source of the biological scaffolding of the linguistic system is the brain itself. That this view is often supposed to brace the biological scaffolding of the linguistic system has already been touched upon in the earlier chapter. This chapter will scrutinize this matter in more detail so that the fallacious assumptions along with the latent orthodoxies can all be jettisoned.

2.2 Neurobiological Foundations of Language and Cognition

One of the best ways of understanding the biological grounding of language and cognitive processes is to look inside the brain to hunt for traces of language in the neural substrate. The reason is, after all, that when we produce and comprehend language, all that we do and engage in can be traced to the activities within the brain. It is thus believed that different kinds of processes that make the whole of language processing viable can be found right in the neurobiological processes on which cognitive processes ride and which drive all of language processing. That this assumption makes for certain apparently naïve but deeply entrenched fallacious routes to slippery lines of reasoning can be exposed by careful examination of the standard methods of neurobiological investigations into language processing.

2.2.1 Brain Imaging Studies

Since we cannot cut open the brain and see what is really going on when we process language, we require certain techniques that do not involve the physical penetration into the brain but at the same time can help us identify and locate the relevant brain activities that can be matched with the tasks that are under our investigation. The temporal sequence of brain activities is best captured by electroneurophysiological studies which are a kind of noninvasive methods of tracing brain activities. There are two methods in electrophysiological studies: One is the electroencephalographic method (EEG) and the other is the magnetoencephalographic method (MEG). While the former works by measuring electrical activities of the brain at the scalp over which some electrodes are placed, the latter works by means of the estimation of the data gathered from the magnetic fields induced over the brain which help determine the location of electrical activities in the brain. Both methods tap the time-locked millisecond-by-millisecond activities of postsynaptic activities across brain cells in a neuronal population. What is crucial here is that it is the signal that is captured by these techniques from which an inference is made about the *source* of the signal as well as about the linguistic process the source in question is an indication of.

It is quite well known that there are certain electrophysiological signatures of different kinds of linguistic processing which can be distinguished by the forms of the linguistic components of a linguistic system which afford the analysis of language processing in terms of its subprocesses. For example, an early lexical-semantic processing anchored within a discourse is detected by the elicitation of the neuro-electrical response which is known as N400 (see Van Berkum et al. 2003), and a negativity response known as ELAN and another response known as P600 are signatures of early and later stages of syntactic processing, respectively (Canesco-Gonzalez 2000; Van Turennout et al. 2003; Friederici 2004). The symbol 'N' (as in 'N400') denotes a negative going wave, whereas the symbol 'P' signifies a positive going wave. Importantly, Baggio

(2018) has argued that N400 reflects a 'cascade' of processes that start with lexical access and end with a kind of top-down binding of features of meaning in a sentential context. Note that the numerical symbol in N400 or P600 reflects the number of milliseconds that it takes for the respective response to be evoked after the given stimulus is registered by the brain. N400 is evoked in cases of lexical selection or access as well as in cases of semantic anomaly, whereas ELAN and P600 are strongly observed in syntactic anomalies and violations of grammatical rules.

Now it is particularly important to note that these electroneurophysiological studies provide no evidence whatsoever about the (neuro)biological grounding of language. All they index is the time flow of particular events in the brain which are correlated with certain parameters of the tasks concerned. For instance, the elicitation of a specific ERP (event-related potential) component (such as N400 or P600) on a certain linguistic task indicates that the relevant neuronal groups within a given brain are sensitive to the specific linguistic task in question *within a specific time window*. From the fact that certain neural activation patterns are sensitive to certain specific linguistic tasks that are marked by the distinct forms of the linguistic components involved, it does not simply follow that the specific properties of the linguistic system that are reflected in the respective types of language processing (lexical properties in lexical access and selection, for instance) are instantiated in the brain processes the signature of which is captured in different ERP components. The difficulty or complexity or the presence/absence of a certain feature in a task is thought to be indexed by a particular ERP response in the brain. This is the underlying logic that drives most electroneurophysiological studies. It is thus easy to notice that the underlying logic is always *defeasible*. If a certain ERP response is an index of a particular parameter of a linguistic task, it does not entail that we have achieved an understanding of the causal connection between the parameter in question and the relevant brain response. All that we gather from this is that there is some causal connection between the task parameter concerned and the relevant brain response. But in the absence of any detailed understanding of the observed causal pattern, it certainly is not safer to conclude that the obtaining of the observed effects of a task parameter is necessarily due to the brain activities as recorded in an ERP response. After all, this could be a case of mere correlations which look like causal connections. Note that this problem applies quite generally to standard cases of indexical relationships as well. Just because smoke is an index of fire, one cannot conclude on the basis of this observation, however frequent it may be, that the causal connection between smoke and fire is understood *just because* the moment has been repeatedly observed. That is, any indexical relationship between two events or objects does not guarantee an accurate understanding of the causal pattern involved.

The other acute problem with electroneurophysiological studies is that the signals from small-scale activities of neuronal cells at postsynaptic junctures are added up to calculate the value of electrophysiological responses of the brain to certain kinds of stimuli of concern. What is patently missing here is the mechanistic linking that can map the electrophysiological patterns to the cognitive activities and processes which are often representationally characterized. To give a simple example, if P600 marks the brain's response to certain syntactic violations, it may be believed that some syntactic relations of linguistic structures are established in the brain at a certain point of time—which is exactly what the P600 component indexes. But this does not tell us anything about the neuro-cognitive mechanisms that link the establishment of the relevant syntactic relations to the electrophysiological response constituting P600. All this tells us is that certain neurological responses occur at certain time intervals. That is all we can gather from this. In other words, the dimension of temporal organization of certain cognitive events may have nothing to do with the organization of mechanisms underlying an event in all other dimensions (such as space and depth). Besides, it is also hard to state how we can shift from the small-scale activities of neuronal cells at postsynaptic junctures constituting electrophysiological responses to the higher-scale constitutional relations of cognitive phenomena (but see Baggio et al. 2014).

Another more general problem with electroneurophysiological studies in particular is the *inverse problem* known in formal and empirical sciences. The inverse problem arises when one draws certain logically valid inferences from the data to the parameters of the object of study which cannot directly be penetrated into. This is called the *inverse problem* because under other normal circumstances one usually draws valid inferences from the parameters of the object of study to the data gathered upon a series of observations—in the inverse problem one goes in the inverse direction in relating data and the parameters of the object of study to one another. For example, we make certain valid inferences about the atmosphere, temperature, and radiations on the surface of the sun by gathering data from outside as we cannot land on the sun to directly measure certain physical values of interest in terms of any parameters we choose. While gathering data from the magnetic fields or electrodes placed over the scalp, we do the same thing as the data from the signals received in the magnetic fields or the electrodes are amassed to measure activation states of the brain along the temporal trajectory of processing in response to certain patterns or kinds of stimuli. The problem is that the same data can be mapped to many possible parameters of activation states of the brain along the same temporal trajectory of processing. Finding out the correct parameter of activation states of the brain is not a trivial problem. Thus, just because the electroneurophysiological data gathered are calculated by averaging out a number of ERP responses of the brain, this does not establish that we have tapped the right brain state that *actually* corresponds to what the brain has been exposed to. It is hard to say, no matter how precise the mapping is, how the same electroneurophysiological data may have been leveraged across a number of states of the brain which can be described in terms of potentially myriad parameters. On the other hand, if one starts by fixing the parameter of brain state within which the electroneurophysiological data will be made sense of, the inverse problem can be mitigated. That is, if one fixes the activation states of the brain (high activations vs. low activations, for example) beforehand so that this suffices to constrain the domain over which the electroneurophysiological data will be collected, one may expect to map the exact form of brain activation states for a given cognitive phenomenon to the ERP components. But this is something we cannot do since identifying and isolating the activation states of the brain cannot be done directly by looking inside the brain. The considerations discussed above apply well to noninvasive brain stimulation methods such as transcranial magnetic stimulation (TMS) where during the performance of a given linguistic task causal conclusions are drawn with respect to the contribution of the stimulated area to a specific brain function (Hartwigsen 2015).

If the temporal flow of neural events is best captured by electroneurophysiological studies, the spatial layout of brain activations is better gauged by certain other imaging methods such as functional magnetic resonance imaging (fMRI) scan or positron emission tomography (PET) scan. fMRI is a technique that traces brain activations by means of the detection of hemoglobin in the blood supply in the brain, given that more activated areas will have a greater flow of oxygenated blood supply. fMRI as a brain imaging technique in essence relies on the emission of radio frequency pulses by the atoms in different neural tissues induced by the magnetic fields placed over the brain. The PET scan technique, on the other hand, consists in capturing the signatures of glucose consumptions by neuronal cells induced by some radioactive tracers administered to the subjects, and these signatures of glucose consumptions by neuronal cells make for the images of the activated areas in the brain. The problems with these studies, especially in connection with the brain–language relationships, are two-pronged. First, both these studies gather information from the lower scales of neuronal organization from molecules and neuronal cells, and from this, the inference that is made is that certain brain areas are activated in certain linguistic tasks and some others are less activated than others. From this, it does not certainly follow that the relevant linguistic task is implemented in the areas that are (more) activated as observed during imaging studies. Worse than this is the fact that this does not also entail that the relevant linguistic property that characterizes the component involved in the linguistic task(s) is instantiated in the brain areas (more) activated. Such a conclusion would be too hasty mainly because the inferential shift from the activity of some neurons to certain representational properties of the linguistic system or to the linguistic task itself on the basis of gathered information from atomic and molecular interactions in the brain is arguably absurd and misleading (see also Johnson 2009, 2012).

Let's take one relevant example in this context which can illustrate this matter more illuminatingly. Thus, we may look over some recent brain imaging studies that have found that certain areas of neural networks (especially Broca's region, parts of basal ganglia in sub-cortical structures and other regions along the superior temporal sulcus) are naturally predisposed for and saliently responsive or sensitive to the central principles of natural language grammar such as hierarchical (rather than linear) composition of linguistic structures (Moro 2008, 2016; Pallier et al. 2011). In this regard, what Moro (2008) says in the context of an inquiry into the neurobiological nature of the essential properties of natural language (such as the hierarchical form of syntax) is representative of what this book considers misguided.

> In fact, in the same way as impossible sentences have led to significant advancement in the understanding of the formal mechanisms behind linguistic competence, impossible languages may lead to significant advancement in *the understanding of the neurobiological nature of the limits of variation across languages*. (emphasis added, p. 161)
>
> When it comes to grammar, structure-dependent syntax is not just one of the possible softwares that this hardware, the human brain, may run, rather *it turns out to be the only software that it construes*--making the hardware-software metaphor itself meaningless when it comes to the relationship between the physical brain and syntax. (emphasis added, p. 186)

This appears to show that the fundamental building principles of natural language grammar are finally implemented or grounded in the neurological constitution of language. But this projects a specious appearance of the neurological instantiation of fundamental principles of natural language grammar. Just because certain neural networks are selectively activated for linguistic structures exhibiting the property P but not for linguistic structures without P, it does not follow from this that P manifested in natural language grammar is itself instantiated in the given neural networks. One reason is that this does not in any way demonstrate that the relevant neural networks (or their selective activation) *but not* others are necessary for the manifestation of P to take place (Rugg 2000). A similar line of reasoning can also be found in Hickok (2014), although the discussion therein is in the context of a critique of the assumed role mirror neurons are supposed to play in cycles of action and perception. Besides, what is also noteworthy is that no neural networks can be said to possess the signature of permanently entrenched functions or perpetually encoded functional abstractions (see Anderson 2014). Hence even those brain areas or neural networks usually activated in response to the hierarchical composition of linguistic structures cannot be said to have been etched with the function or property of responding to the structural hierarchy of linguistic structures. What such studies at most show is that the relevant neural networks are selectively attuned to the hierarchical composition of linguistic structures in virtue of the organizational functional sensitivity of the neural networks involved. This is a modest claim, but to claim more than this is ungrounded. To provide one helpful analogy, one may consider tropisms in plants. When plants are saliently sensitive to light or gravity for the growth of certain parts exposed to the stimuli of light or gravity, one cannot comfortably assert that the properties of light or gravitational properties are directly instantiated in plant cells or tissues. Salient activation of neural networks in response to hierarchical rules of linguistic structures cannot smoothly be translated into the neurobiological instantiation of the property of hierarchical constitution of linguistic structures. That certain neurobiological structures are saliently sensitive to hierarchical rules of linguistic structures and are thus markedly activated more often in response to such inputs is not any more surprising than the fact that certain neurological structures (V1, V2, V4 areas) in the visual cortex are saliently attuned to the properties of color.

Second, in many imaging studies some calculations are made in order to arrive at some apparently valid conclusions about brain–language relationships. Correspondences between the activated neurological locations and certain linguistic tasks are brought forward to make claims about the brain networks subserving the installation of particular linguistic components and/or functions. This is often informed by a background theory of neurological functions and a model of language. In most cases, the brain signal data from a controlled task are subtracted from the brain signals that underlie the experimental condition for a linguistic task (say, reading some well-formed words or sentences in a language). Likewise, many signals for linguistic tasks in a sequence are also aggregated to yield the overall value of particular brain activation parameters. The deeper problem with such analyses is that such additive or subtractive calculations of brain signals for an understanding of brain–language relationships may be profoundly misguided (see also Van Lancker Sidtis 2006). It is quite plausible that the representational properties of the linguistic tasks involved are such that they linger on even during controlled tasks the brain signals of which are subtracted. Plus the way the brain processes certain parts of linguistic tasks may well be non-additive—that is, the operations do not necessarily add up to something that can contribute to some aggregate values. Rather, the trajectories of the brain functions involved may form a lattice-like network of activation patterns for a given sequence of tasks the brain is exposed to (see, for discussion, Uttal 2013). This impugns the role of imaging studies in revealing how the linguistic system may be (neuro) biologically instantiated.

2.2.2 Lesion Studies, Language Disorders, and Other Neurological Cases

If brain imaging studies do not provide unequivocal evidence concerning the neurobiological instantiation of the linguistic system whose components are reflected in various aspects of language processing, one may now insist that the right path to understanding the nature of the neurobiological instantiation of the linguistic system lies in the neurological origins of language disorders that stem from brain injuries, strokes, or brain lesions caused by accidents or otherwise. The classic cases of such disorders are Broca's aphasia and Wernicke's aphasia both of which result from brain damages or strokes. The specific names of the two kinds of aphasia derive from the names of the neurologists who discovered them first, that is, Paul Broca and Carl Wernicke. While Broca's aphasia is an expressive disorder—which implies that the patient in Broca's aphasia cannot express grammatically well-formed sentences or phrases, whereas Wernicke's aphasia is a receptive disorder in which language understanding is impaired giving rise to the production of linguistic expressions which may appear syntactically well formed but have no intelligible meanings (see, for a classic discussion, Luria 1966). Thus, it seems that Broca's aphasia and Wernicke's aphasia are just polar opposites in that language production and language comprehension can be dissociable and so disrupted. Although most cases of aphasia are described in terms of *modes* of language processing—language production vs. language comprehension or fluency vs. non-fluency, only recently have researchers attempted to relate the specific nature of a particular case of aphasia to its underlying linguistic component (Van Lancker Sidtis 2006). For instance, Broca's aphasia is associated with a disruption of the network that subserves the syntactic and (morpho)phonological components of the linguistic system, whereas Wernicke's aphasia is believed to attach to disruptions of brain functions for the conceptual and lexical-semantic components of language (Caplan 1996). However, the associations are not always readily determinable and neatly distinguishable. For instance, Broca's aphasia is not univocally a disturbance in the syntactic component because language comprehension is also disrupted in patients of Broca's aphasia, suggesting that the semantic system is also, at least in part, disrupted (Caramazza et al. 2005; Neuhaus and Penke 2008; Murray 2018; but see Grodzinsky 2000). On the other hand, Wernicke's aphasia is not also a deficit that exclusively affects the lexical-semantic and conceptual systems since errors in syntactic representations are also found in Wernicke's aphasics (Faroqi-Shah and Thompson 2003; Caplan et al. 2004). The demarcation problems with both Broca's aphasia and Wernicke's aphasia thus remain unresolved (see Fava 2002; Basso 2003).

The underlying logic of most aphasic studies is that a particular brain function or a set of brain functions is supposed to be subserved by a certain brain area, especially when that specific brain area is damaged, and consequently, the observed behavior in the victim who has undergone the brain damage shows signs of poor performance in the individuated brain function(s) that the damaged area presumably executes. The problem with this apparently reasonable assumption is that it conflates two different kinds of relations routed from the same source. One is the relation between the location of a lesion and the particular syndrome that (supposedly) results from the lesion, and the other is the relation between the neurological locus of the lesion and the particular brain function(s) that the brain area (supposedly) carries out. Notice that both these relations are derived or routed from the same entity, that is, the specific brain locus under investigation. As has been argued at the beginning of this chapter, on the one hand, similar lesions within the closer boundaries of the same brain locus may produce distinct profiles of aphasia across individuals, and the identification of a given type of aphasia or any other syndrome does not necessarily provide an unequivocal indication of the exact brain locus (e.g., conduction aphasia which is supposed to result from damage to brain areas that connect Broca's and Wernicke's areas and exhibits some features of the profile manifested in Wernicke's aphasia cannot be restricted to any particular neurological locus). On the other hand, the same brain locus affected during a kind of aphasia or any other syndrome does not always produce the same (unique) functional disturbance in any linguistic ability, and at the same time, it is not equally true that any unique functional role can be attributed to the same brain locus that is found to be affected for the same type of aphasia. That this is so is because in many cases the patterns of errors in a given linguistic ability observed in a patient often depend *not on* the neuronal groups and pathways that participate in carrying out the linguistic ability concerned *but rather* on those neuronal groups and pathways that do not perform the functions involved in that particular linguistic ability (Geschwind 1984; also see Dick and Tremblay 2012; Rutten 2017).

If the considerations laid out above point to something worth taking into account, it is misleading to invariably attribute the functional impairment in a certain linguistic task that involves some component(s) of the linguistic system to that specific brain region which has been found to be damaged. As far as the role of the neurobiological substrate vis-à-vis cognitive functions is concerned, the manifestation of any cognitive function depends on something, while the emergence of errors in any cognitive function originates in something else. These two notions—the manifestation of any cognitive function and the emergence of errors in any cognitive function—are not the same thing as they are often mistakenly conflated. The manifestation of any cognitive function rests not merely on the mediation by the neurobiological substrate but also on the representational properties of the system that constitute the cognitive function concerned, whereas the emergence of errors in any cognitive function turns on a lot more factors other than those solely restricted to the cognitive function at hand that go beyond relations of neurological implementation with respect to the cognitive function under consideration. To put it in other words, the former more often than not exclusively hinges on considerations that pertain to the cognitive function at hand, while the latter seldom if ever springs from factors that belong to the domain boundaries of the particular cognitive function in question. The reason is that the emergence of errors in any cognitive function in any case of syndrome after a certain kind of brain damage is never due to the fact that only the cognitive function concerned is disrupted. Rather, it may well be that some other related cognitive functions or the supporting cognitive mechanisms that have a different cognitive constitution from the one under consideration are actually at fault. It is really difficult to pinpoint any single factor when errors in any cognitive function consequent to some brain damage emerge.

The convoluted nature of the relation between cognitive functions for language and neurobiological loci does not, after all, make a good case for the instantiation of the linguistic system in the neurobiological substrate. One may now insist that for something to be neurobiologically instantiated, it is not necessary that what is instantiated will have a one-to-one mapping with the neurobiological substrate in which that thing is instantiated. For, there could be many levels between the level of implementation and the level of what is implemented that clutter the scenario, by rendering invalid the optimal condition for a one-to-one mapping. Even if we grant that this is the case, there is no denying that such a scenario provides no evidence whatever about the neurobiological instantiation of the linguistic system. Just because some connections between the damage to certain brain regions and the impairment of the corresponding cognitive functions are observed, we cannot necessarily conclude that the cognitive system of which these functions are a part is actually implemented in the brain regions or even in the neuronal networks that are disrupted due to the brain damage. For all we know, this certainly does not establish that we are actually dealing with some neurobiological mechanism(s) that is(are) found to be disrupted and hence have led to the disappearance or impairment of some cognitive function. That is, this in no way points to any neurobiological mechanisms which, one may believe, mediate the cognitive function impaired. Plus damages to the brain (i.e., brain lesions) *ultimately* affect the elements existing at the lower scale of neural organization. That is to say that what is actually affected in lesions is not any brain region per se; rather, it is the neuronal cells, assemblies of neurons, neuronal networks that are affected or damaged. By studying these parts of brain organization, one cannot make any conclusive inference about the neurobiological mechanism(s) that can realize a given cognitive function, let alone precisely specify the intermediate mechanisms that link the neurobiological stuff to the cognitive structures.

The deeper problem this raises is that one cannot circumvent the hurdles that arise from this issue by simply stating that brain activations in cases of brain damages can tell us something about the cognitive capacity concerned as we match the nature of brain activations with the observed performance. There is no unambiguous way in which one can simply interpret more or less activation in any neurological locus and then attach the level of brain activations to some facet of the cognitive system whose function has been disrupted. On one interpretation, more activation in a neurological locus may imply that the cognitive function under consideration is intact. But on another interpretation, it could also mean that the person at hand is facing difficulty with the cognitive function because the cognitive function is now compromised or affected. Likewise, less activation in a neurological locus may indicate either that the cognitive function at hand is compromised or that the cognitive function has been absorbed into the brain mechanisms and thus hardened into an automatic procedure (see also Van Lancker Sidtis 2006). The observed performance or behavior of the person examined in a given linguistic task cannot thus be properly aligned with the neurological signatures that we find out. There is no standard solution to the fundamental conundrums this matter gives rise to. One has to appraise this deeply in order to be able to understand how attempts to secure the biological foundations of language and cognition in the brain fail.

There are also other issues of concern that need attention at this juncture. What is presupposed in most (neuro)biological studies on language and cognition is that language is biologically unique and hence language and other non-linguistic cognitive capacities must be segregated in biological terms. What this implies is that while one examines the biological foundations of language, one actually investigates the nature of the capacity for language in various domains of linguistic components but not aspects of visual processing tasks or motor activities of a non-linguistic nature (such as drawing). This means that language and other cognitive domains are *biologically* segregated from one another, and this biological segregation can be adduced as a fact to provide scaffolding for the supposition that the linguistic system is ultimately anchored in our (neuro)biological substrate. Closer scrutiny reveals that this supposition is deeply problematic.

First, the biological segregation of language from other non-linguistic domains (such as vision, memory, motor system, thinking and reasoning, and emotions) is not entirely a matter of biological separation in kind, although it seems that it is. The reason is that language is distinct from vision or memory not simply because vision or memory involves neural structures that are distinct from those involved in language, or because they involve distinct genomic patterns of expression. There is no way one could actually demonstrate that language, vision, memory, motor system, etc., are distinct systems that are biologically differentiated in a transparent manner in the brain or in gene expressions. This is because these cognitive systems or the facets of these abilities are determined at the outset from outside *by us*. One cannot pre-conceptually (or pre-theoretically) isolate or fix the boundaries of neural structures and genomic expressions and then map them to disparate cognitive systems such as language, vision, and memory. In every case, we first determine what the cognitive system is and demarcate the boundaries of the system by characterizing the properties and aspects of that system so that the biological foundations of the system in question can be traced to only those biological structures that are sensitive to or capable of manipulating the manifestation of parts pertaining to the system concerned. In a nutshell, no one who studies the biological foundations of language and cognition puts the cart before the horse in actual practice.

Second, studies that make too much of cases of biologically rooted dissociations between language and other cognitive capacities carry certain flawed assumptions that need to be unpacked. One of the underpinning ideas in many neurolinguistic studies is the concept of *double dissociation* which means that if two cognitive abilities, say A and B, can be independently impaired or spared, then the two abilities must be biologically distinct. For instance, SLI is a case which is believed to provide evidence for the situation in which language is impaired, but other non-linguistic abilities are spared. On the other hand, Williams Syndrome or Turner's Syndrome is believed to present the converse situation, that is, a situation in which language is spared but other cognitive capacities are impaired. To consider one example, Turner's Syndrome is a genetic disorder that occurs in females and in which most linguistic abilities (producing grammatically well-formed sentences, accurate comprehension, reading, writing, etc.) are preserved but other visual and spatial capabilities are severely disrupted. A child with Turner's Syndrome cannot draw or copy a picture, build or assemble any structure out of blocks, etc. Similar arguments are also made about the dissociation between linguistic and non-linguistic visual-spatial capabilities in brain-damaged sign language users. For example, left-hemisphere damaged signers with the resulting sign language aphasia have been found to construct correct descriptions of visual-spatial configurations through their aphasic sign utterances, whereas right-hemisphere damaged non-aphasic signers produce in accurate sign language utterances visual-spatial descriptions that indicate neglect of the left side of space (see, for relevant discussion, Curtiss 2013). From these observations, it is certainly easy for one to jump to conclusions, but the postulated explanation is far from true.

The fundamental problem with such arguments is that the *grain* size of double dissociations is never made explicit. When one examines two cognitive capacities A and B and finds that A is impaired but B is intact, or conversely, that B is impaired but A is preserved, one does not know how much of A is impaired (or preserved) *and* how much of B is preserved (or impaired) that can enable the generalization that A and B can be doubly dissociated. It is radiantly clear that in most cases the grain size is rather coarse such that some amount of impairment in one cognitive capacity is aligned with a great deal of preservation of other cognitive capacities in order to make a case for double dissociations. If, on the other hand, the grain size for comparison is made finer, many of the apparent disparities between one cognitive capacity and certain other cognitive capacities cashed out in terms of heterogeneous manifestations may disappear (see also Karmiloff-Smith 1992; Juola and Plunkett

2000; Dunn and Kirsner 2003; also see Davies 2010). Furthermore, the argument for double dissociations does not display the other side of the picture. That is, when one observes that A is impaired but B is intact, or conversely, that B is impaired but A is preserved, facets of dissimilarities between A and B along the dimensions of preservation and impairment are highlighted. What is left out is the picture which might show the ways in which some aspects of *both* A and B (together) are either preserved or impaired. This is a kind of dishonesty that masks the real scenario, thereby authorizing the scholars to make extravagant claims about (biological) *modularity* of cognitive systems supported by double dissociations. Thus, in no sense do the supposed double dissociations provide the justification required for the claim that the linguistic system or linguistic cognition is (neuro)biologically instantiated. What cases of supposed double dissociations at least show is that some aspects of processing in one cognitive capacity can be independent of processing in another cognitive capacity. Even if the argument for double dissociations were cogent and adequately accurate in the way the scholars involved believe that it is, this would not establish that the linguistic system or linguistic cognition is (neuro)biologically instantiated, for all cases of supposed double dissociations avoid furnishing an account of the underlying biological mechanisms that are relatively impaired or preserved as one makes a transition from one cognitive capacity to another. For double dissociations to make biological sense, one has to provide evidence of those neuro-cognitive mechanisms (linking biological mechanisms to the cognitive system upward) which are disrupted (or preserved) in one cognitive capacity but intact (or disrupted) in another. Simply speaking, all cases of supposed double dissociations bypass levels of analysis and also analyses of levels.

Third, studies of linguistic savants seem to be another fertile ground for explorations into the nature of the (neuro)biological instantiation of linguistic cognition. The argument for supposed double dissociations reappears here in another apparently stronger form. Linguistic savants are individuals who possess magnified linguistic capabilities with or without deficits in other cognitive capacities. The case of Laura, a child who was found to produce and comprehend complex clauses with remarkable spontaneity even though her abilities in counting, arithmetic, drawing, and also auditory recognition were largely compromised, is one such example. What was, however, evident is that Laura's syntactic-morphological abilities were far ahead of her semantic-pragmatic abilities in language (Yamada 1990). Smith and Tsimpli (1995) present another such case which is of a boy called Christopher, whose linguistic competence was found to include the partial mastery of 16 languages from different language families, although he was not able to perform everyday routine visual and motor activities (drawing, buttoning clothes, etc.). These cases are believed to show how language and the rest of cognition can be rooted in separable biological mechanisms. On the face of it, this sounds like a reasonable and plausible description or explanation of the observed patterns in linguistic savants. But again, the arguments marshaled turn out to be specious upon closer inspection. One problem is that if linguistic cognition and non-linguistic cognition are biologically segregated in these cases, it appears that non-linguistic cognition may not be required for supporting linguistic cognition or vice versa. But the question is how one marks out the boundaries of non-linguistic cognition that can be shown to be used or not used for supporting linguistic cognition. It is rather doubtful or at least not clear that Laura or Christopher had no cognitive capacities that could have structured or supported their linguistic cognition as it was manifested in them (see also Bates 1997). How much of the amount of non-linguistic cognition should one deduct from what is normally manifested in humans in order to arrive at the conclusion that non-linguistic cognition can be disrupted and yet linguistic cognition can be operative? For all we know about language–cognition relations, this question is not easy to answer.

Similarly, how much of the amount of knowledge in one component of language (say, syntax) can be compared to the amount of knowledge in another component of language (say, semantics or pragmatics) so that one could assert that a person's linguistic ability is magnified in one linguistic component but compromised in another? No matter how adequate a quantitative analysis of linguistic abilities in linguistic savants is, this cannot in any way show where to make the right cut so that the elevation of linguistic competence over non-linguistic capacities in such individuals can be adequately and unambiguously demonstrated (see also Shatz 1994). Thus studies of linguistic savants neither show that double dissociations between language and other cognitive systems exist nor establish that the linguistic system or linguistic cognition recognized as such is biologically instantiated. The augmented linguistic abilities in linguistic savants may well be due to unexamined and hitherto unknown neuro-cognitive mechanisms that magnify as well as augment just those subsets of cognitive capacities that support the linguistic machinery. What is observed in linguistic savants is that some (or at least some facets of) other cognitive capabilities are compromised. This does not at all show that these linguistic savants have *no* cognitive capacities preserved, or even that *all* cognitive capacities that can support language are impaired. Nothing of this sort is established.

In all, biological foundations of language and linguistic cognition do not univocally indicate that the language system is implemented in our neurobiological infrastructure. What it overall shows is that many facets of linguistic cognition have a neurobiological imprint which can be better understood in terms of the functional architecture of the neuronal organization. However, this does not imply that the neurobiological imprint of certain aspects of linguistic cognition is equivalent to neurobiological instantiation of the linguistic system. These are two different things. In many cases, it becomes clearer that the neurobiological description of patterns of linguistic phenomena is a re-description of the phenomena in question garbed in a pseudo-mechanistic neurobiological language. This particular problem is in essence similar to the problem of leaving out the mechanistic account of psycholinguistic mechanisms which come clothed in a pseudo-computational re-description of aspects of the linguistic components involved (Seidenberg 1988). In more general terms, a pseudo-computational re-description of certain representational characteristics of the linguistic system which purports to make reference to the neurobiological scale is as good (or bad) as a pseudo-mechanistic neurobiological re-description of functional marks of neurolinguistic cases. What is remarkably common between the two approaches is the trail of elusions of the neurobiological mechanisms when a relation between the neurobiological organization and the corresponding cognitive architecture is drawn. The only difference lies in the placement of the direction of the postulated type of explanation. In motivating psycholinguistic accounts of the linguistic system, one may come up with the formulation of a functional architecture for a given aspect of some component of the linguistic system (phonological priming, for example) with a view to having it finally integrated seamlessly into the neurobiological organization. The direction of explanatory fit is from the neurobiological organization to its cognitive plausibility, whereas relating the cognitive profile of any neurolinguistic case (whether it is a disorder or some other trait) to the neurobiological base projects downward to the neurobiological mechanisms from the functional characterization of the relevant linguistic deficit or trait.

It is vital to recognize that the two different descriptions of linguistic cognition meet one another exactly at the converging point which occupies the zone where some unknown unexplored neuro-cognitive mechanisms are believed to be operative. The assumption is, of course, vindicated by the particular transductions of cognitive structures into neural structures and vice versa the trails of which are somehow palpably observed but barely understood. This problem has been appraised in certain circles as being related to the hard problem of consciousness since the mental-to-neural reduction of linguistic phenomena is confronted with the problem of how to deal with first-person accounts of linguistic constructs and concepts (see Kowalewski 2017). What is important is that the point made just above presumably motivates accounts of the cognitive with models of representations, activities, and processes at varying levels of detail always shifting their positions within the nested layers of scales that encompass both the cognitive and the neurobiological. This does not mean that the models that attempt to tame the problem of tightening the strings that can link the cognitive to the neurobiological are in themselves flawed. Rather, the projection of these models is distorted because the projection loses a lot of detail when the projection (re)encodes what is actually coded in transitions from the cognitive to the neurobiological or vice versa.

2.3 How May Linguistic Cognition Ride on Neurobiology?

The fact that language has several components and dimensions makes it harder to apply the right sort of concept of language as it may be relevant to neurobiology. On the one hand, the internal components of language are its syntactic, semantic, morphological, and phonological systems along with the lexicon that partakes of all the systems mentioned. But, on the other hand, there is a conception of language as a code which is understood as enveloping human language as a whole. On this conception, natural language is conceived of as a phenomenon or as a capacity with its external manifestation in speech utterances/expressions, words, writing, and also in the ability to express concepts and experiences. These two are distinct *dimensions* of language since the former relates to the internal structures, constituents, and resources of natural language, whereas the latter is linked to the way language manifests itself in the organism. When one speaks of the neurobiological instantiation of the texture of linguistic cognition, it is the linguistic capacity enveloping human language as a holistic system that is governed by networks of gene expressions, pathways of neural networks, formation of neural assemblies and associations (see Friederici 2017). Perhaps relevant to this is the distinction between the language-ready brain and the very linguistic code itself (see Hillert 2015) because the linguistic code as a varying phenomenon or capacity can fit into whatever the nature or form the language-ready brain assumes at any stage of brain evolution. The biologically constrained growth of human language targets those parts of brain wiring that result in the kind of structuring of neurobiological networks responsible for the manifestation of the linguistic capacity. Here, matters of constitution of phrases, principles of word order, semantic, and morpho-phonological composition, etc., are not laid out or specified by gene expressions or developed brain structures. Rather, once the capacity for language is instituted by the constraints of biological processes, this paves the way for the expression or manifestation of the internal parts of the linguistic system. This happens much like the way the visual capacity after being instituted by way of the development of eyes, visual pathways, and regions in the brain gives rise to the visual processing routines and centers for various kinds of representation of objects, surfaces, scenes, and events as part of visual cognition. But the key difference in the current context is that the vision-ready brain directly expresses the visual system generating visual cognition, whereas the language-ready brain does not directly express the linguistic system that manifests linguistic cognition. The reason this is so is that there are (and have been) distinct natural languages but distinct tokens of vision do not exist. While it is understandable that both language and vision are bio-cultural hybrids in the sense that cultural/environmental/ecological influences often (re)organize the visual system as in the case of the augmentation of the visual capacity in, say, bird watchers, car experts, etc., the degrees of freedom that the visual system can exercise in this regard are quite limited. Hence, the amount of variation vision or, for that matter, olfaction can have is bounded by the nature of the hybridity vision or olfaction allows for. Visual cognition does not thus vary very widely across the perceptual parameters of shape, size, texture, motion, color, etc., whereas languages do vary quite widely in terms of sounds, meanings, syntactic constructions, and, most importantly, words. The cultural/environmental/ecological influences in the nature of hybridity manifest in human language are far greater than those for vision or olfaction. That is why it takes some years (usually 5–6) for babies to acquire a human language, whereas human vision under normal circumstances does not require years to develop.

Therefore, the linguistic capacity may be invariant across cultures or populations of humans, but aspects of linguistic cognition as part of the manifestation of the linguistic system may not be so. Thus, what is instantiated in language-ready brain is the linguistic capacity *but not* the linguistic code or system itself that manifests or evinces linguistic cognition. A failure to mark this distinction begets confusions about the way language rides on neurobiology. The linguistic system or the texture of linguistic cognition is not directly expressed by the language-ready brain because its internal parts and components need not be structured by any biological processes or structures. Thus, when the internal parts and components of the linguistic system (in terms of syntactic, semantic and phonological processing) are observed to be realized in neurobiological structures (in brain imaging or electrophysiology, or through lesions, for example), it does not follow that the system *itself* is inherently located in, or structured by, the relevant neural pathways, networks, and associations activated for language. What this shows is that the linguistic system or linguistic cognition displays certain properties which are processed by certain centers and pathways of the brain. Even when neurobiological accounts make fairly specific claims apparently about decomposed aspects of the linguistic capacity such as the ones for semantic structuring, inflectional morphology, or syntactic encoding, these accounts unfailingly refer to and capture the dynamic (i.e., online) processing of aspects of semantic anomaly, syntactic encoding, morphological compositions, etc. Insofar as such accounts cover exclusively the processing of the components of the complex linguistic system, these neurobiological descriptions do not actually isolate one component of the linguistic capacity from another by way of decomposition. A neurobiological account of syntactic encoding, for example, does not isolate the syntactic capacity from the linguistic capacity, for this would imply that the semantic capacity, the phonological capacity, the morphological capacity, and so on as isolable components go on to make up a monolithic linguistic capacity. For this to obtain, genetic and evolutionary constraints will have to be fine-grained enough to make such distinctions for component parts of the linguistic capacity in humans—something that biology does not allow for either in development or in evolution (see West-Eberhard 2003; Carroll 2008). As stressed above, what happens in such cases is precisely that the components of the complex linguistic system, *but not of the linguistic capacity*, exhibit properties which are processed upon evocation by certain centers and pathways of the brain. An analogy might be of help here. Just because numbers are processed in the brain with certain centers being activated for numerical cognition (brain structures including the left angular gyrus), to say that the system of number itself is instantiated in the brain is at best misleading.

The role of (neuro)biological constraints is thus pivotal to the emergence and continuation or preservation of the linguistic capacity as a whole in the lifetime of humans. When neuroscientific methodologies are deployed to examine and probe into certain aspects of the linguistic system, it is not the linguistic capacity per se that is now under scrutiny. Rather, it is the properties of the linguistic system that become manifest after the emergence of the linguistic capacity which are tracked by different neuroscientific methodologies that penetrate into the brain through invasive or noninvasive techniques. Biological constraints are a precondition for the emergence of the linguistic capacity, and the emergence of the linguistic capacity is a precondition for the manifestation of the facets of the linguistic system. When the facets of the linguistic system become manifest, all that remains left is how the biological machinery (including the brain) within which the linguistic capacity is instituted works with the components of the linguistic system. Interestingly, a description of the relevant distinction is also succinctly sketched out in Bickerton (2014) below.

> We may not know much about the brain, but we know that when it comes to the structure of sentences, a derivational model is much closer to the kind of thing brains routinely do than is a representational model. (p. 38)

Here, the 'representation model' would consist of all the aspects and properties of the linguistic system that arise after the emergence of the linguistic capacity as a whole. But, on the other hand, the 'derivational model' would include everything about the way the brain in virtue of being situated in the organism as a whole performs its functions with those aspects and properties of the linguistic system. This is much like the way a printing machine, for example, after being constituted by all operational parts functions with different kinds of items that go in the making of printed items. The way the machine thus performs its operations on such items designates the derivational model, whereas the properties or aspects of the printed objects thus yielded refer to the representational model. In the present context, neurobiological descriptions of online syntactic or semantic or even phonological processing point to the nature and form of the operations characterized by the brain's working with components of the linguistic system, *but not* to the linguistic capacity itself. That is because the linguistic capacity is not something that can be tracked by (neuro)biological techniques of investigation—all that can be or is actually tracked is what remains left after the linguistic capacity is already in place. The linguistic capacity is a mode of description but not a system having determinable parts or components.

We may thus wonder if there is a way in which language–biology relations can be marked in demarcating terms. One way of thinking about bridging the gap in language–biology relations is to adopt the three-level schema of Marr (1982) and then locate whatever is linguistically structured at the computational level with the algorithmic and implementational levels instantiating the psychological mechanisms and neurobiological mechanisms, respectively. This is in line with what Embick and Poeppel (2014) seem to think. Thus, we can also hope that the more we discover something about the psychological mechanisms and then relate them gradually to the underlying neurobiological mechanisms for any given linguistic phenomenon, the prospects of integration of the computational level with the algorithmic and implementational levels are quite strong. The actual bet can be placed on the algorithmic level because this is what can be believed to furnish a description of the bridging mechanisms (which may be deemed to be psycholinguistic) needed for the integration. But the problem at this stage is that even if the integration of the computational level with the algorithmic and implementational levels is achieved for quite a lot of linguistic phenomena across and within languages, this cannot guarantee in any way prospects of the linguistic being unified with the neurobiological. This is so because reduction does not necessarily coincide with integration and unification, especially when the theory that reduces one description to another may well be a mere medium of encoding (i.e., a kind of re-description of that which is to be integrated with another description) but not necessarily a level of bridging principles (see Bechtel 1986). Besides, even if integration does project bridging principles, this does not in any way postulate or warrant the unity of the linguistic with the neurobiological at all explanatorily significant levels or scales that may be deemed sufficient, as discussed in Chapter 1.

Therefore, it seems wiser to acknowledge that there are, and must always be, gaps in language–biology relations that can be regarded as cases of disharmony no matter how unity by means of integration or otherwise is achieved at whatever levels one takes into consideration for any given arena of inquiry. Hence, a rich understanding of the linguistic organization of, rather than mere contribution to, cognition is more pivotal to studies of the mind. This does not, however, distance linguistics from the cognitive sciences; rather, this helps linguistics to *unequivocally* balance its autonomous parts with those parts affording the benefits of integration with other cognitive sciences, notably neuroscience, in understanding the nature of language–cognition relations. This can suitably and selectively enrich our view of cognition vis-à-vis the natural world of life processes because the more we go into the biological realm, patterns that are unmatched surprise us and then we seek to place them in the category of biological uniqueness. But any questions of biological uniqueness have to be judged against their relevance to the entity whose uniqueness is at issue. Just as what is cognitively bounded cannot dictate what there is or could be in biology, biology cannot dictate what there exists in cognition. Much more on this will be said in Chapter 5.

2.4 Summary

This chapter has attempted to show that the biological foundations of language and language–cognition relations are rife with conundrums that are far deeper than is generally assumed among most researchers who delve into the connections obtaining between our biological infrastructure and linguistic cognition. As has been discussed in the sections above, the particular problems and troubles arise mainly from certain misunderstandings of the exact role of biological mechanisms that may mediate the manifestation of linguistic cognition. Even if some parts of the problems involved in making aspects of linguistic cognition biologically viable for implementation can be tamed, the strategies employed do not at all render linguistic cognition biologically transparent. On the one hand, no genetic study on linguistic cognition via the route taken through language disorders lends itself to something that can be considered to contribute to the genetic instantiation either of linguistic cognition or of the linguistic system recognized as such. On the other hand, neurobiological studies on language and language disorders tend to bypass the neurobiological, so much so that nothing substantive is displayed that can link the neurobiological to the functional/cognitive. Ultimately, this ends up exaggerating either the neurobiological on certain occasions or the functional/cognitive on certain others. Such amplification is neither warranted nor justified by the mission that an account of the biological instantiation of linguistic cognition or the linguistic system may be supposed to accomplish. Given these ineluctable dilemmas, we may now try to figure out how on earth language and cognition relate to one another. This is what we shall turn to now as Chapter 3 aims to make explorations into very nature of the relation of derivation within the fabric of language–cognition relations.

References

Anderson, M. L. (2014). *After Phrenology: Neural Reuse and the Interactive Brain*. Cambridge: MIT Press.
Baggio, G. (2018). *Meaning in the Brain*. Cambridge: MIT Press.
Baggio, G., van Lambalgen, M., & Hagoort, P. (2014). Logic as Marr's computational level: Four case studies. *Topics in Cognitive Science, 7*, 1–12.
Balaban, E. (2006). Cognitive developmental biology: History, process and fortune's wheel. *Cognition, 101*(2), 298–332.
Basso, A. (2003). *Aphasia and Its Theory*. New York: Oxford University Press.
Bates, E. (1997). On language savants and the structure of the mind: Review of *The Mind of a Savant: Language Learning and Modularity*. *International Journal of Bilingualism, 1*(2), 163–179.
Bechtel, W. (1986). The nature of scientific integration. In W. Bechtel (Ed.), *Integrating Scientific Disciplines* (pp. 3–52). Dordrecht: Springer.
Bickerton, D. (2014). *More Than Nature Needs: Language, Mind and Evolution*. Cambridge, MA: Harvard University Press.
Bishop, D. V. M. (2006). What causes specific language impairment in children? *Current Directions in Psychological Science, 15*(5), 217–221.
Brown, C. M., & Hagoort, P. (Eds.). (2000). *The Neurocognition of Language*. New York: Oxford University Press.
Canesco-Gonzalez, E. (2000). Using the recording of event-related brain potentials in the study of sentence processing. In Y. Grodzinsky, L. Shapiro, & D. Swinney (Eds.), *Language and the Brain: Representation and Processing* (pp. 229–267). New York: Academic Press.

Caplan, D. (1984). The mental organ for language. In D. Caplan, A. R. Lecours, & A. Smith (Eds.), *Biological Perspectives on Language* (pp. 8–30). Cambridge: MIT Press.

Caplan, D. (1996). *Language: Structure, Processing and Disorders*. Cambridge: MIT Press.

Caplan, D., Waters, G., DeDe, G., Michaud, J., & Reddy, A. (2004). A study of syntactic processing in aphasia I: Behavioral (psycholinguistic) aspects. *Brain and Language, 91*, 64–65.

Caramazza, A., Capasso, R., Capitani, E., & Miceli, G. (2005). Patterns of comprehension performance in agrammatic Broca's aphasia: A test of the Trace Deletion Hypothesis. *Brain and Language, 94*, 43–53.

Carroll, S. (2008). Evo-Devo and an expanding evolutionary synthesis: A genetic theory of morphological evolution. *Cell, 134*, 25–36.

Curtiss, S. (1977). *Genie: A Linguistic Study of a Modern-Day 'Wild Child'*. New York: Academic Press.

Curtiss, S. (1994). Language as a cognitive system: Its independence and selective vulnerability. In C. Otero (Ed.), *Noam Chomsky: Critical Assessments* (pp. 211–255). London: Routledge.

Curtiss, S. (2013). Revisiting modularity: Using language as a window to the mind. In M. Piattelli-Palmarini & R. C. Berwick (Eds.), *Rich Languages from Poor Inputs* (pp. 68–90). New York: Oxford University Press.

Davies, M. (2010). Double dissociation: Understanding its role in cognitive neuropsychology. *Mind and Language, 25*(5), 500–540.

Dick, A. S., & Tremblay, P. (2012). Beyond the arcuate fasciculus: Consensus and controversy in the connectional anatomy of language. *Brain, 135*(12), 3529–3550.

Dunn, J. C., & Kirsner, K. (2003). What can we infer from double dissociations? *Cortex, 39*, 1–7.

Embick, D., & Poeppel, D. (2014). Towards a computational(ist) neurobiology of language: Correlational, integrated and explanatory neurolinguistics. *Language, Cognition and Neuroscience, 30*(4), 357–366.

Enard, W., et al. (2002). Molecular evolution of FOXP2, a gene involved in speech and language. *Nature, 418*, 869–872.

Evans, V. (2014). *The Language Myth: Why Language Is Not an Instinct*. Cambridge: Cambridge University Press.

Faroqi-Shah, Y., & Thompson, C. K. (2003). Effect of lexical cues on the production of active and passive sentences in Broca's and Wernicke's aphasia. *Brain and Language, 85*(3), 409–426.

Fava, E. (Ed.). (2002). *Clinical Linguistics: Theory and Application in Speech Pathology and Therapy*. Amsterdam: John Benjamins.

Feldman, J. A. (2006). *From Molecule to Metaphor: A Neural Theory of Language*. Cambridge: MIT Press.

Fisher, S. E. (2006). Tangled webs: Tracing the connections between genes and cognition. *Cognition, 101*(2), 270–297.

Fisher, S. E., & Vernes, S. C. (2015). Genetics and the language sciences. *Annual Review of Linguistics, 1,* 289–310.

Friederici, A. D. (2004). Event-related brain potential studies in language. *Current Neurology and Neuroscience Reports, 4*(6), 466–470.

Friederici, A. D. (2017). *Language in Our Brain: The Origins of a Uniquely Human Capacity*. Cambridge: MIT Press.

Friedmann, N. & Rosou, D. (2015). Critical period for first language: The crucial role of language input during the first year of life. *Current Opinion on Neurobiology, 35,* 27–34.

Gallistel, C. R., & King, A. P. (2009). *Memory and the Computational Brain: Why Cognitive Science Will Transform Neuroscience*. New York: Wiley-Blackwell.

Geschwind, N. (1984). Neural mechanisms, aphasia and theories of language. In D. Caplan, A. R. Lecours, & A. Smith (Eds.), *Biological Perspectives on Language* (pp. 31–39). Cambridge: MIT Press.

Gopnik, M. (1990). Genetic basis of grammar defect. *Nature, 346*(6281), 226.

Gopnik, M., Dalalakis, J., Fukuda, S. E., & Fukuda, S. (1997). The biological basis of language: Familial language impairment. In M. Gopnik (Ed.), *The Inheritance and Innateness of Grammars* (pp. 111–140). New York: Oxford University Press.

Grimaldi, M. (2012). Toward a neural theory of language: Old issues and new perspectives. *Journal of Neurolinguistics, 25*(5), 304–327.

Grodzinsky, Y. (2000). The neurology of syntax: Language use without Broca's area. *Behavioral and Brain Sciences, 23,* 1–21.

Haesler, S., et al. (2004). FoxP2 expression in avian vocal learners and non-learners. *Journal of Neuroscience, 24,* 3164–3175.

Hagoort, P. (2009). Reflections on the neurobiology of syntax. In D. Bickerton & E. Szathmáry (Eds.), *Biological Foundations and Origin of Syntax* (pp. 279–298). Cambridge: MIT Press.

Hagoort, P. (2014). Nodes and networks in the neural architecture for language: Broca's region and beyond. *Current Opinion in Neurobiology, 28,* 136–141.

Hartshorne, J. K., Tenenbaum, J. B., & Pinker, S. (2018). A critical period for second language acquisition: Evidence from 2/3 million English speakers. *Cognition, 177*, 263–277.
Hartwigsen, G. (2015). The neurophysiology of language: Insights from non-invasive brain stimulation in the healthy human brain. *Brain and Language, 148*, 81–94.
Herschensohn, J. (2007). *Language Development and Age*. Cambridge: Cambridge University Press.
Hickok, G. (2014). *The Myth of Mirror Neurons: The Real Neuroscience of Communication and Cognition*. New York: Norton.
Hillert, D. G. (2015). On the evolving biology of language. *Frontiers in Psychology, 6*, 1796.
Hurst, J. A., Baraitser, M., Auger, E., Graham, F., & Norel, S. V. (1990). An extended family with a dominantly inherited speech disorder. *Developmental Medicine and Child Neurology, 32*, 352–355.
Johnson, G. (2009). Mechanisms and functional brain areas. *Minds and Machines, 19*, 255–271.
Johnson, G. (2012). The relationship between psychological capacities and neurobiological activities. *European Journal for Philosophy of Science, 2*(3), 453–480.
Johnston, J. R. (1997). Specific language impairment, cognition and the biological basis of language. In M. Gopnik (Ed.), *The Inheritance and Innateness of Grammars* (pp. 161–180). New York: Oxford University Press.
Juola, P., & Plunkett, K. (2000). Why double dissociations don't mean much. In G. Cohen, R. A. Johnston, & K. Plunkett (Eds.), *Exploring Cognition: Damaged Brains and Neural Networks—Readings in Cognitive Neuropsychology and Connectionist Modeling* (pp. 319–327). Sussex: Psychology Press.
Kaan, E. (2009). Fundamental syntactic phenomena and their putative relation to the brain. In D. Bickerton & E. Szathmáry (Eds.), *Biological Foundations and Origin of Syntax* (pp. 117–134). Cambridge: MIT Press.
Kang, C., & Drayna, D. (2011). Genetics of speech and language disorders. *Annual Review of Genomics and Human Genetics, 12*, 145–164.
Karmiloff-Smith, A. (1992). *Beyond Modularity: A Developmental Perspective on Cognitive Science*. Cambridge: MIT Press.
Katsos, N., Roqueta, C. A., Estevan, R. A. C., & Cummins, C. (2011). Are children with specific language impairment competent with the pragmatics and logic of quantification? *Cognition, 119*(1), 43–57.

Kowalewski, H. (2017). Why neurolinguistics needs first-person methods. *Language Sciences, 64,* 167–179.

Lenneberg, E. (1967). *Biological Foundations of Language.* New York: Wiley.

Leonard, L. (2014). *Children with Specific Language Impairment.* Cambridge: MIT Press.

Levelt, W. J. M. (2008). *Formal Grammars in Linguistics and Psycholinguistics.* Amsterdam: John Benjamins.

Lewontin, R. (2000). *The Triple Helix: Gene, Organism and Environment.* Cambridge, MA: Harvard University Press.

Luria, A. R. (1966). *Higher Cortical Functions in Man.* New York: Basic Books.

MacDermot, K. D., et al. (2005). Identification of FOXP2 truncation as a novel cause of developmental speech and language deficits. *American Journal of Human Genetics, 76*(6), 1074–1080.

Marr, D. (1982). *Vision: A Computational Investigation into the Human Representation and Processing of Visual Information.* San Francisco: W. H. Freeman.

Marshall, J. C. (1980). On the biology of language acquisition. In D. Caplan (Ed.), *Biological Studies of Mental Processes* (pp. 106–148). Cambridge: MIT Press.

Mondal, P. (2014). *Language, Mind, and Computation.* London: Palgrave Macmillan.

Moro, A. (2008). *The Boundaries of Babel: The Brain and the Enigma of Impossible Languages.* Cambridge: MIT Press.

Moro, A. (2016). *Impossible Languages.* Cambridge: MIT Press.

Murray, L. L. (2018). Sentence processing in aphasia: An examination of material-specific and general cognitive factors. *Journal of Neurolinguistics, 48,* 26–46.

Neuhaus, E., & Penke, M. (2008). Production and comprehension of *wh*-questions in German Broca's aphasia. *Journal of Neurolinguistics, 21*(2), 150–176.

Newbury, D. F., & Monaco, A. P. (2010). Genetic advances in the study of speech and language disorders. *Neuron, 68*(2–13), 309–320.

Newport, E. L., Bavelier, D., & Neville, H. J. (2001). Critical thinking about critical periods: Perspectives on a critical period for language acquisition. In E. Dupoux (Ed.), *Language, Brain and Cognitive Development* (pp. 481–502). Cambridge: MIT Press.

Pallier, C., Devauchelle, A.-D., & Dehaene, S. (2011). Cortical representation of the constituent structure of sentences. *Proceedings of the National Academy of Sciences USA, 108,* 2522–2527.

Pedersen, N. L., Plomin, R., & McClearn, G. E. (1994). Is there G beyond g? (Is there genetic influence on specific cognitive abilities independent of genetic influence on general cognitive ability?). *Intelligence, 18,* 133–143.

Piattelli-Palmarini, M., & Berwick, R. C. (Eds.). (2012). *Rich Languages from Poor Inputs.* New York: Oxford University Press.

Plomin, R. (2011). Why are children in the same family so different? Non-shared environment three decades later. *International Journal of Epidemiology, 40*(3), 582–592.

Rugg, M. D. (2000). Functional neuroimaging in cognitive neuroscience. In C. M. Brown & P. Hagoort (Eds.), *The Neurocognition of Language* (pp. 15–36). New York: Oxford University Press.

Rutten, G. (2017). *The Broca-Wernicke Doctrine: A Historical and Clinical Perspective on Localization of Language Functions.* Berlin: Springer.

Sampson, G. (2005). *The 'Language Instinct' Debate.* London: Continuum.

Seidenberg, M. S. (1988). Cognitive neuropsychology of language: The state of the art. *Cognitive Neuropsychology, 5*(4), 403–426.

Sesardic, N. (2005). *Making Sense of Heritability.* Cambridge: Cambridge University Press.

Shatz, M. (1994). Review of Laura: A case for the modularity of language. *Language, 70*(4), 789–796.

Smith, N., & Tsimpli, I. (1995). *The Mind of a Savant: Language Learning and Modularity.* Oxford: Blackwell.

Sprouse, J., & Hornstein, N. (2016). Syntax and the cognitive neuroscience of syntactic structure building. In G. Hickok & S. A. Small (Eds.), *Neurobiology of Language* (pp. 165–174). Amsterdam: Elsevier.

Stromswold, K. (2001). The heritability of language: A review and meta-analysis of twin, adoption and linkage studies. *Language, 77,* 647–723.

Stromswold, K. (2006). Why aren't identical twins linguistically identical? Genetic, perinatal and postnatal factors. *Cognition, 101,* 333–384.

Stromswold, K. (2008). The genetics of speech and language disorders. *New England Journal of Medicine, 359*(22), 2381–2383.

Uttal, W. R. (2013). *Reliability in Cognitive Neuroscience: A Meta-Meta-Analysis.* Cambridge: MIT Press.

Van Berkum, J. J. A., Zwitserlood, P., Hagoort, P., & Brown, C. M. (2003). When and how do listeners relate a sentence to the wider discourse? Evidence from the N400 effect. *Cognitive Brain Research, 17*(3), 701–718.

van der Lely, H. K. J., & Christian, V. (2000). Lexical word formation in children with grammatical SLI: A grammar-specific versus an input-processing deficit? *Cognition, 75*(1), 33–63.

Van Lancker Sidtis, D. (2006). Does functional neuroimaging solve the questions of neurolinguistics? *Brain and Language, 98,* 276–290.

Van Turennout, M., Schmitt, B., & Hagoort, P. (2003). When words come to mind: Electrophysiological insights on the time course of speaking and understanding words. In N. O. Schiller & A. S. Meyer (Eds.), *Phonetics and Phonology in Language Comprehension and Production: Differences and Similarities* (pp. 241–278). Berlin: Mouton de Gruyter.

Vargha-Khadem, F., et al. (1995). Praxic and nonverbal cognitive deficits in a large family with a genetically transmitted speech and language disorder. *Proceedings of the National Academy of Sciences, 92,* 930–933.

Watkins, K. E., Dronkers, N. F., & Vargha-Khadem, F. (2002). Behavioural analysis of an inherited speech and language disorder: Comparison with acquired aphasia. *Brain, 125*(3), 452–464.

West-Eberhard, J. (2003). *Developmental Plasticity and Evolution.* New York: Oxford University Press.

Yamada, J. E. (1990). *Laura: A Case for the Modularity of Language.* Cambridge: MIT Press.

3

Cognition from Language or Language from Cognition?

Thus far, we have examined ways in which language or linguistic cognition can be said to be biologically instantiated. Two specific types of biological instantiation with respect to the nature of the linguistic system or linguistic cognition have been specially considered. Closer inspection reveals that there is in fact no straightforward and non-roundabout way of having the system of language biologically instantiated. The problems that arise from such a sort of biological instantiation, if that becomes a brute-force choice for one to make, eat into the very system of biological organization of cognitive phenomena because the requisite relations do not simply obtain. Given that these problems and dilemmas make one rethink the nature of linguistic cognition as it does not somehow fit into the overall scheme of instantiation relations within the biological infrastructure, we may perhaps understand things better by stepping into the territory of linguistic cognition. But linguistic cognition recognized as the level of cognitive organization that emerges from the linguistic system available to human being does not seem to lend itself to being immediately disclosed in such a manner that we can simply tell the linguistic portion apart from the cognitive portion or even vice versa while we look inside linguistic cognition. In simpler terms, the question here revolves around the issue of whether linguistic cognition springs from language as we understand it or from the general format of cognition as reflected in our cognitive capacities and abilities that do not naturally pertain to language. For all we know, this question is not trivial and simple at all, for there are deep theoretical and conceptual commitments involved which thwart attempts at teasing the truth out. In many ways, the overarching conundrums in deriving the general format of cognition from language or in distilling linguistic cognition into the general structure of cognition are more often than not a matter of our own descriptions of cognition and language supported by the conceptual and theoretical machinery we have constructed. Needless to say, rooting through the maze of conceptual and theoretical cross-connections for the appropriately demarcated relation between language and cognition appears to be an unpalatable exercise. Nevertheless, in this chapter we shall undertake to deal with this question since the risk involved is worth the fruits that we may gain in making our preparatory conceptualization enriched enough to approximate to the formulations to be developed in the next chapter.

If one contends that the whole of cognition or at least the most part of cognition derives from language, one does not merely endorse a view of the primacy of language over the structure of cognition. This view may carry with its concomitant assumptions about the ontogenetic development and evolution of language and cognition. Thus, the basic format of cognition can be said to have been structured by language through the pathways along which language and cognition have evolved in the past or generally develop. This does not, of course, prevent one from adopting the stance that the properties and aspects of language as a cognitive domain, partially or otherwise, constitute the infrastructure of cognition as we understand it in its *synchronic* sense. This amounts to saying that language as a cognitive system structures or modifies the system of cognition, and there can be various ways in which the structuring or modifying role of language in cognition can be conceptualized. Similar lines of reasoning apply equally well to the view that the fundamental architecture of language derives from the structure of cognition. This is concordant with the statement that the central properties and facets of language accepted either as an abstract system or as an operational cognitive domain are distilled from the basic organization of cognition. Thus, the central properties and facets of language form a subset of the set of properties and facets of cognition. Given these two possibilities, one may be compelled to choose one view or other and can also expect to gather the required evidence that will possibly weigh with either of the stances. But, as a matter of fact, the situation is not as black and white as it looks. In this chapter, we shall see that the pieces of evidence show that both possibilities may be simultaneously valid, given that reasonably unambiguous characterizations of both language and cognition are at hand. We shall first deal with the issue of how the fundamental architecture of language derives from the organizational principles of cognition and then move on to take up the question of whether the basic format of cognition owes its essential character to language.

3.1 Language from Cognition

The natural question that concerns us now is whether the fundamental architecture of language derives from the organizational principles of cognition. And if this really obtains in natural contexts, we may also be inclined to know how such a derivation can get off the ground. Understanding this question may equip us with the necessary conceptual paraphernalia that can be easily extrapolated to the obverse side of the relation between language and cognition, that is, determining whether the basic format of cognition is due to language. In a nutshell, the aim is to locate and pin down the signatures of cognitive organization in language. If the nature of language is such that it can project a window onto the cognitive architecture, the underlying supposition is that facets of cognitive organization are somehow crystallized and thus integrated into the fabric of language. Insofar as this can be figured out by examining the structures of natural language, one may believe that language mirrors the structure of cognition. For instance, the following sentences uncover some interesting signatures of our cognitive organization.

(11) ?The film depicts in a manner that is both elegant and emotionally satisfying with varying kaleidoscopic patterns the scene.
(12) ?She did not intend to call the meeting which everyone believed would change the trajectory of the future we all hoped to see off.
(13) ?We want to convey that if there is anything wrong it must be the idea that those who lag behind are to be scorned and trashed to children.
(14) ?I have no idea who these teachers that have blighted the lives of many students are.

The sentence in (11) demonstrates that particular phrases that are actually intrinsic parts of certain predicates must be contiguous (not merely linearly, but hierarchically) with the predicates concerned, especially if there is some intervening material that has a complex structure (involving a relative clause, for instance) disturbing the structural and semantic dependency between the predicates and the phrases that constitute the conceptually intrinsic parts of those predicates.[1] That is, the arguments of predicates must be placed structurally close to the predicates if some complex modifying expression comes between the predicates and the arguments. A workable notion of contiguity or closeness that we can adopt here (in a framework-neutral sense) is in offer: if X is attached in a phrase Y along with Z (when X and Y are sister nodes) or if X is attached in a phrase Y with no further embedding or addition of V within X, then X and Z are contiguous. Here, X, Y, Z, and V are all phrases. Note that the first part of this formulation captures

---

[1] It is not always the case that any intervening material between a predicate and its argument can disrupt the syntactic and semantic dependency between the predicate and some argument of the predicate. In some cases, the presence of an intervening material does not simply render a sentence unacceptable. Rather, it can make it more straightforward and explicit, as the following pair taken from Larson (1989) shows.

(i) You see large numbers of Dr. Who fans at such conventions.
(ii) You see at such conventions large numbers of Dr. Who fans.

Here, the intervening material 'at such conventions' in (ii) does not disturb the dependency between the predicate 'see' and its argument 'large numbers of Dr. Who fans.' But important to note is the fact that the intervening material here is not so complex after all, precisely because it does not involve a complex item such as a relative clause or an entire clause as is the case in (11–14).

3 Cognition from Language or Language from Cognition?   101

hierarchically attached phrases that are contiguous, while the second part introduced after 'or' subsumes linearly attached expressions that are contiguous. Also, notice that this formulation allows for degrees of closeness or contiguity. In any case, what is significant in the present context is that arguments of predicates are conceptually intrinsic parts of predicates in the sense that the meaning of a predicate remains incomplete if the linguistic roles of its arguments are not specified. Thus, in (11) the noun phrase 'the scene' is the (second) argument of the predicate 'depict,' and the dependency between 'the scene' and 'depict' is best expressed if it is not structurally disturbed by some complex modifying expression. Since the modifying expression 'in a manner that is both elegant and emotionally satisfying with varying kaleidoscopic patterns' comes before 'the scene,' it disrupts this dependency, thereby rendering the sentence unacceptable. The sentences (12–13), on the other hand, show that expressions that are not, strictly speaking, arguments of predicates must also be adjacent to the predicates if the arguments of these predicates form complex intervening expressions. Thus, the particle 'off' is part of the phrasal verb 'call off' meaning something different from the verb 'call' and hence cannot remain far removed from it when the argument of the predicate 'call off' is complex enough (as in 'the meeting which everyone believed would change the trajectory of the future we all hoped to see'). The same point can be raised about the relation between the predicate 'convey' and the modifier 'to children.' There is a parallel between (11) and (14) in that in (14) the dependency between the predicate 'are' and 'these teachers' is disturbed by the complex modifying expression 'that have blighted the lives of many students.'

What these examples clearly demonstrate is that the conceptual connection or relation between a predicate and its arguments (which are conceptually intrinsic elements of predicates) or modifiers (which are conceptually optional elements of predicates) must be realized in a linguistic structure in a contiguous fashion if and only if some complex expression may be placed between them. Even if this principle is integrated and built into the fabric of linguistic structures, this is in no way a linguistic principle per se. The reason is that the *structural* disturbance in establishing a conceptual connection or relation between a predicate and its arguments or modifiers derives from the *cognitive* disruption attaching to the indexing of conceptual structures or entities in our (short-term or intermediate-term) memory which prefers immediate resolutions of all dependency links. If some complex expression intervenes between the predicate and its arguments or modifiers, the memory resources are allocated to the task of determining the dependency relations within the intervening complex expression—which results in a kind of fading or gradual weakening in the initial indexing and chaining of the conceptual structures or entities initiated by the very appearance of the first predicate. The parsing machinery works much like a finite-state automaton which operates on the basis of transitions that are modulated by the closest preceding states in moving to the successive states. Any expression that comes packaged with its own ensemble of dependency links involving various conceptual structures or entities engages and occupies the space of the parsing machinery to the full, thereby giving rise to the deterioration of linkages that flank the complex intervening expression in question. Overall, this indicates that some principles of our cognitive organization are somehow embedded and frozen in the very texture of grammar. In this sense, linguistic structures can be reckoned to have absorbed the imprints that our cognitive machinery imposes on language. There is, in fact, more to this matter than meets the eye.

There is, in fact, another sense in which language mirrors the operational constraints of the cognitive system. The growth of linguistic structures complies with the restrictions that are inherent in the cognitive system. The growth of linguistic structures takes place by virtue of concatenation of smaller pieces of linguistic structure, and this incremental growth of concatenation depth is an index of the complexity of linguistic structures constructed. This growth can actually take place in three different portions—the right side, the center, and the left side. The following sentences exemplify the three cases, respectively.

> (15) This book is so interesting that it evokes feelings of awe which one would not naturally feel in the presence of someone who is always ready to step into a territory that is not a fertile ground for explorations into the unknown (which) ...

3   Cognition from Language or Language from Cognition?   103

(16) The huge building the architect everybody ... loves looks after overlooks the river.

(17) This woman who has always bothered to tend the children who do not receive the support that they need from those who ought to be close to the children to help them in a way which is justifiably desirable for parents (who) ... is very sincere.

The dots in each of the sentences indicate that the construction may go on indefinitely. What is interesting here to note is that the incremental growth of concatenation in any such case manifests some formal properties which emanate from the facets of the cognitive apparatus that underlies language processing. First of all, no predicate in natural language can have more than three arguments—which implies that the cognitive machinery that has to do the processing of linguistic structures cannot index more than three individual objects which can become the arguments of a predicate. In English, for instance, we reach the maximum number in predicates such as 'give' or 'put' which requires three participants (one agent who does the act of giving, one recipient and the thing given in the case of 'give,' and one agent, the thing put and the location where the thing is put in the case of 'put'). Given that this is the case, it is not immediately obvious that the incremental growth of concatenation of pieces of linguistic structures must never carry forward dependencies by taking all the arguments of a given predicate. That is to say that it is not clear why the linguistic growth of concatenation cannot pick out all the arguments of a given predicate already present in the structure and then create dependencies that connect all the arguments chosen from the previous structures. This can be illustrated by taking into account (16), for example. Notice that the sentence in (16) says that the architect looks after the huge building and, similarly, everybody loves that architect—which clearly tells us that in the first level of concatenation of a relative clause the second argument (which is the object) of the predicate 'look after' is 'the huge building' (which is the first argument of the predicate 'overlook'), the first argument of 'look after' (which is the subject) being taken over by 'the architect,' and then the second argument of the predicate 'love' is shared with the first argument of 'look after' (i.e., 'the architect') in the next level of concatenation and so on. Each level of concatenation depth carries forward only some, but not all, of the arguments of a given predicate in a linguistic construction that is unfolding. Similar considerations apply to (15) and (17) and any other conceivable cases of incremental concatenation of pieces of linguistic structures. This generalization can be schematized the following way.

(18)

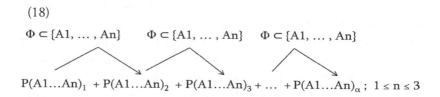

$$\Phi \subset \{A1, \ldots, An\} \quad \Phi \subset \{A1, \ldots, An\} \quad \Phi \subset \{A1, \ldots, An\}$$

$$P(A1\ldots An)_1 + P(A1\ldots An)_2 + P(A1\ldots An)_3 + \ldots + P(A1\ldots An)_\alpha \, ; \, 1 \leq n \leq 3$$

The formulation in (18) states that a series of predicates, that is, Ps, with each containing some arguments A1…An, can form a sequence determined by the nature of concatenation of linguistic structures in such a way any level of concatenation must carry forward only a subset Φ of the actual number of arguments in any P in the series. Now, this generalization does not concern the representational properties of language per se, but rather the character of our cognitive machinery that cannot handle too many dependency chains that can be created from all the available arguments of a predicate.

Language parallels our cognitive organization in many other ways. Thus, both the imprints of the workings of our cognitive machinery and distinct forms of cognitive organization which can be regarded as structural patterns of cognition are reflected in language. This is particularly evident in the conceptual infrastructure of language in which the structure of thought is reflected. For example, the functions of thought involve differentiation, categorization, mnemonic realization, analysis, analogy making, pattern matching, etc. Thought functions in a way that makes such cognitive configurations lead to appropriate behaviors/actions which turn on intentions on the part of agents and the concomitant or consequent interactions of cognitive processes with the world as a function of states of dynamics where complex equations go on to determine their reliability, predicted effects, state transitions, etc. Aside from that, the epistemic function of thought appears to be implanted in language—language helps internalize, represent, manipulate, and transform the structure of knowledge as it is embedded in human beings who think and reason and take actions on the basis of such thinking and reasoning (Clark 1997, 2008; Johnson 2018). If language can be said to enhance the epistemic potential and capacity of the species, the covert supposition is that language has somehow inherited the cognitive organization of thoughts and thinking, thereby strengthening its intrinsic link to thoughts and incorporating in itself (at least some of) the epistemic properties of thoughts. It is also, no doubt, true that our society places special emphasis on epistemic confidence with the consequence that failure in tasks evincing epistemic confidence triggers and meets with disappointment, ignorance, dismissal, dismay, and even insult (Hirstein 2005). Therefore, there is no denying the fact that we have gradually become epistemic vehicles.

In a more general sense, language represents things, abstractions, and lots of many other entities by expressing them in a format determined by linguistic forms within and across languages. Assuming that language is a second-order system that formats sensory-perceptual-motor representations which are transduced into linguistic representations via a number of mapping channels, we can say that it is thought as part of the general fabric of cognition which constitutes one of these mapping channels feeding sensory-perceptual-motor representations into language. Hence language has borrowed some facets of its own representational character from thoughts and thinking, given that thought *in itself* is an ensemble of representational processes that define the macrostructure of cognition which typically encodes concepts, patterns of construal, associations of ideas, etc. It is now easy to note that this presupposes that thoughts and thinking can be cast in a variety of media one of which is language, and also that the representational or rather intentional functions of thought and language may vary because of their informational functions. This is indeed the line of reasoning that has been employed by some people (see Millikan 2005, 2018). The crucial point Millikan makes is that thought does not always need language and vice versa, because she believes they have as parallel systems different types of intentionality. This implies that the intentionality of language is different from that of thought, on the grounds that the intentionality of language resides in *how* or in what *manner* the linguistic forms serve their linguistic functions by means of their forms, while the intentionality of thought relates to the way intentional attitudes (such as thinking, believing, desiring, proposing, promising, feeling) serve their functions of correlating with structures in the outer world. These conventional functions of cognitive capacities are also termed *proper functions*. In this connection, Millikan also emphasizes that these proper functions, which are akin to biological functions (the function of the heart, for example) in many respects, have no direct connection to individual human intentions or thoughts. Rather, they have a connection to language users' utterances which can have a generally descriptive character, given that utterances are to be understood in a generic sense.

In this sense, what seems at least plausible is that language, by way of inheriting the macrostructure of thoughts and thinking, implements representational processes needed for desirably optimal actions in appropriate environments which in some way or another contribute to *epistemic systematization* and *synchronization* in our species. Epistemic systematization involves the formation of new knowledge by using stored knowledge structures, either endogenously or as a function of pressures from the environment, in ways of keeping track of new knowledge through proper arrangement, whereas epistemic synchronization refers to the construction of new knowledge out of the existing structures by staying in tune with them and making the new structures available for further manipulation in synchrony with the existing ones. While the former helps in the categorization and codification of knowledge structures, the latter facilitates the use of these structures in real time (say, in planning, understanding, reasoning, judging, etc.).

At this stage, one may be apt to point out that language is in its epistemic functions a reflex of thought and thinking is due to the ontogenetic and phylogenetic trajectories that connect language, thought and thinking together. Thus, it does not seem surprising that the development of language in children is observed to parallel the emergence of the capacity to understand other human beings' minds as human children begin to start considering other human beings around them to be intentional from around the ninth month (Tomasello 1999, 2019). And from about the fifth year onwards, children gradually exhibit an understanding of others' perspectives, that is, they go their way to fully develop a 'theory of mind' (Gopnik and Astington 2000; de Villiers 2007). This demonstrates that language or its emergence evinces and thereby imports the representational systems of the mental machinery. But the character of thinking being restructured by language links language itself to the sensitivity of thinking to being molded by language. That is to say that unless thinking contains the potential for being restructured inherent in it, language cannot transform thinking. In this way, thinking too modulates how language can shape thinking, for it is not the case that every kind of thinking (say, tactile thinking) is so affected by language. Vygotsky (1962) arrived at an illuminating conclusion when he showed that initially in the first few months, thought and language develop in the children almost separately, but as soon as language takes a more mature shape in the ontogenetic pathway; it somehow influences the growth of both private and cultural thinking through a kind of 'private speech.' This sort of private speech also helps configure cognitive niches through the manipulation of representational properties in the brain, as Vygotsky has argued. Likewise, it is not hard to imagine that language assumed its representational character by piggybacking on whatever cognitive structures were earlier available in our ancestors as rudimentary forms of thought for expression were plausibly present in our ancestor primates (Köhler 1927; Tomasello and Call 1997; Hurford 2007). If we assume that thought was already present in the hominids around 6 million years ago or even before, it must have taken some time for language to absorb the properties of thought which are now observed to be built into the linguistic system. Whatever the path of evolution has actually been, it seems clear that language was perhaps initially more like an unstructured form that was soaked in the structures of thought which supported and thus formatted language for good. A similar line of reasoning has been adopted by Bickerton (2014) who has argued that a system of language not fully developed and realized, also called 'protolanguage,' evolved to become the fullblown language we see now, but its form without any externalization by the articulator apparatus constituted thoughts as available through language. Further nuances in thought catalyzed by the system of protolanguage at last reinforced the need for the language faculty, he continues to argue. In all, this points to the parallel lines of the emergence of language and thought which may be brought forward to bolster the point that language itself needed the emergent thought to become what it is now.

Given this understanding of the cognitive imprints in language, one may be tempted to assert that language is not merely a symbolic system that carries cognitive imprints, but it is *intrinsically* a system of cognitive structures. This view is prominently expressed in Cognitive Grammar (Lakoff 1987; Langacker 1987; Talmy 2000), which holds that syntactic-phonological units of language are symbolic units which are mapped onto representations of conceptualizations. What is fundamentally important is that such conceptualizations are thought to be grounded in the sensory-motor, perceptual systems and processes. This indicates that conceptualizations are derived from and ultimately anchored in our cognitive organization constituted by perception, memory, and categorization. Thus, mental representations and processes underlying thoughts and reasoning are exactly the elements that are embodied in the symbolic units of natural language expressions. Hence the direction of fit appears to be from mental processes and thoughts to natural language expressions. In this sense, linguistic expressions can be said to be structured in terms of how they are conceptualized, and this suggests that linguistic expressions are in fact extensions of cognitive structures. Aspects of embodiment pertaining to sensory-motor experiences often determine the range of possible meanings corresponding to those cognitive structures. The following examples illustrate this well.

(19) The man is jumping over the fence.
(20) The man's jumping over the fence is unacceptable.

These two sentences have the same common predicate, that is, 'jump,' which *profiles* (specifies and elaborates on) a process which is viewed in two different ways indicating two different modes of perceptual processing. The event profiled by the predicate 'jump' in (19) is constituted by a series of sub-events which make up the event of jumping, and this sequence of sub-events is conceptualized in such a manner that the preceding sub-events are mapped onto the succeeding sub-events in a sequence. This is how a whole event is mentally tracked in space and time via the series of sub-events that actually constitute the whole event. This is called the *sequential* mode of cognitive organization of an event which occurs in real time. On the other hand, the predicate 'jump' in the sentence (20) profiles a process of a quite different kind designating a distinct mode of cognitive organization. The nonfinite clause 'jumping over the fence' adjoined to the possessive noun phrase 'the man's' does not designate an event as it occurs in real time. Rather, it individuates an event in its whole which is construed in a *summative* fashion in which the individual sub-events are *sort of* frozen, collapsed, and rolled into a single unified mental representation. This is supposed to occur when an event is conceptualized as a gestalt retrieved from the memory. In this sense, a summative mode of conceptualizing an event may be superimposed and thus rides on the sequential mode of conceptualization, on the grounds that it is the sequential mode of conceptualization that constructs schematic representations of a dynamic sequence of sub-events which are transduced into a gestalt-like form for further manipulations. Overall, this shows that natural language expressions are themselves a reflex of our cognitive organization. There are more such cases that serve to illuminate the same point.

(21) Jeet met Preet.
(22) Preet met Jeet.
(23) Jeet and Preet met.

The differences between the sentences in (21–23) can be cashed out in terms of the *natural paths* of cognitively salient elements that form parts of conceptualizations. Natural paths in a clausal structure reflect the ordering of the elements in the structure of a clause which are cognitively natural or salient (Langacker 1987). This implies that some elements in a certain type of structure are more cognitively natural or salient than others in other types of clausal structure, and if this is the case, this is bound to yield differences in the patterns of conceptualization reflected in linguistic structures. What is interesting above is that in the sentence (21) it is 'Jeet' which is cognitively more salient than 'Preet' and hence constitutes what is called the 'trajectory,' while 'Preet' is the not-so-focal entity with respect to the cognitively natural ordering and hence 'Preet' is the 'landmark.' A landmark is the background entity against which the trajectory is supposed to be located, thereby mirroring the structure of perceptual conceptualization in which some entity becomes the focal entity only when viewed against the background set by another entity. The sentence (22) reverses the natural ordering of cognitively salient elements present in (21) since it is 'Preet' rather than 'Jeet' that forms the trajectory against the background set by 'Jeet.' The case in (23) merges these two patterns of conceptualization in such a way that the two different elements that had different natural paths in (21–22) cease to be distinct as both 'Jeet' and 'Preet' in the noun phrase 'Jeet and Preet' come to constitute a single unified trajectory with the landmark being implicit. It is noteworthy that the structures of conceptualization in all the three sentences are not grounded in more fine-grained linguistic structures. Rather, the structures of conceptualization manifest in the sentences are taken to be linguistic correlates of cognitive structures or imprints of cognitive processes.

There is another way in which linguistic structures may be reckoned to reflect facets of cognitive organization and mental representations. The (conceptual) substance of linguistic structures can be regarded as something which is different from linguistic expressions per se in the sense that syntactic and phonological expressions in language can be thought of as formal elements that have correspondences with a distinct system of formal-semantic units the substance of which derives from mental representations. Such an approach is found in Jackendoff's (1983, 2002) theory of Conceptual Semantics within the framework of the broader cognitive approach to meaning. Conceptual Semantics proposes that it is a *conceptual structure* which allows us to connect to the world out there via the world projected within the mind. Hence, conceptual structure is a distinct level of mental structure that encodes the world as human beings conceptualize it. What is unique about conceptual structure is that it is independent of both syntax and phonology which are, or *at least* can be, language-dependent or language-specific systems, but conceptual structure is connected to syntax and phonology by interfaces having interface rules which consist of words, among other things, that connect conceptual structures to syntactic and phonological structures. Conceptual structures, staying at an independent

3 Cognition from Language or Language from Cognition?

level of thought and reasoning, are built in a combinatorial manner out of conceptually distinct ontological categories such as Object, Place, Direction, Time, Action, Event, Manner, Path. Combinatorial structures built out of such categories encode category membership, part–whole relations, predicate-argument structure, etc. In this sense, it seems that conceptual structures as expressed in linguistic structures are the right source of insights into the projection of cognition onto language. The following sentences are indicative of this insight.

(24) These men like Toyota cars.
(25) The girls pulled up Jane on her inappropriate behavior.
(24') [$_{\text{Event object}}$MEN LIKE $_{\text{object}}$CARS($_{\text{property}}$ Toyota)]
(25') [$_{\text{Event object}}$ GIRLS PULLUP (+PAST) $_{\text{object}}$ JANE *cause* BEHAVIOR ($_{\text{object}}$JANE & $_{\text{property}}$ INAPPROPRIATE)

The conceptual structures of (24) and (25) are roughly those in (24') and (25'). These representations are thus supposed to be the mental correlates of the linguistic structures concerned. What is striking in these representations is that some linguistic expressions are themselves used as ingredients of the relevant mental representations. This may raise deeper questions about the metalinguistic circularity, given that some natural language expressions are themselves supposed to be parts of non-linguistic mental structures. Besides, the status of conceptual primitives such as Event, Go, Manner, Path may also be debated on the grounds that these conceptual primitives are linguistically valid symbols only in English and hence may invoke the problem of language-specific semantic bias. Notice that this specific problem does not arise within the context of the articulation of cognitive structures as is found in Lakoff (1987) and Langacker (1987). As a matter of fact, this appears to be a problem especially in view of the goal this section sets out to achieve, that is, to show that linguistic structures incorporate patterns of our underlying cognitive organization. This is more so because the conceptual structures of linguistic expressions are in themselves constituted of linguistic expressions, so much so that it is hard to see how conceptual structures can really achieve independence from the formal elements of linguistic expressions. This requirement becomes a pressing concern when one considers the intrinsic character of conceptual structures accepted as such. In such a case, it is not really clear how linguistic structures can be deemed to reflect cognitive representations and aspects of our cognitive organization since the individuation of the mental structures or of the cognitive correlates concerned is forever deferred to a circularly never-ending chain of events that constitute the decoding of linguistic expressions involved in a given conceptual structure. In the face of this dilemma, one may attempt to postulate more intermediate levels of mapping between a given conceptual structure and the linguistic expression whose conceptual structure it is. Notwithstanding its appeal in formal quarters, this does not serve to isolate the conceptual correlates or the imprints of cognitive organization that are to be detected in linguistic expressions in order that the cognitive makeup of the linguistic system can be succinctly characterized and distinctly grasped.

Thus, it seems that the characterization of mental structures without reference to linguistic expressions that are supposed to designate conceptual primitives has the better prospect of revealing the cognitive character of natural language. Since there can be several possible choices for the exact semantic primitives or ontological categories actually selected, there can also be several corresponding forms of conceptualizations. That is, what is really present in the mind may remain invariant, whereas the structures of conceptualizations will vary depending on the exact semantic primitives or ontological categories chosen. This problem accepted as such, at least in part, emanates from the metalinguistic fallacy inherent in most attempts to characterize the structural form of conceptualizations. Therefore, attempts to describe linguistic meanings or conceptualizations in terms of natural language structures and/or some logical symbols/formulas end up widening the distance between what is actually conceptualized in the mind and the representational devices. Although it must be recognized that this becomes an inevitable choice in certain circumstances, this does really fall short of projecting a window onto the cognitive texture of natural language in unequivocal terms. This does not simply serve the purpose that has been set in the current context.

If the language-respective rendering of mental structures manifest in language vitiates the prospect of having a fine-tuned description of the cognitive texture of linguistic structures, an approach that does away with the medium of language for the expression of cognitive imprints in language may be considered more optimal. The theory of *conceptual spaces* (Gärdenfors 2004, 2014) is just such a step toward this goal, in that conceptual spaces are structured around a *geometrical* organization of certain dimensions that mark the conceptual units of linguistic expressions. Thus, conceptual spaces inherent in linguistic expressions are cast in a spatial medium which is believed to be intrinsic to the very format of concepts and conceptualizations found in language. For instance, the English expression 'The blue car is my favorite machine' will map the conceptual space designated by 'the blue car' to that designated by 'my favorite machine.' The former will contain in its spatial network dimensions of shape, color, size, value, function, and the latter will have dimensions such as weight, size, function, shape, value. The mapping from the former to the latter must connect the dimensions such as function and value in the corresponding conceptual spaces. Similarly, verbs such as 'push,' 'pull' designate action patterns that constitute action vectors in an action domain which may be called the *force domain*. Since vectors have magnitude and direction, they can participate in all formal operations that are typical of vectors. A set of vectors that can be realized within a particular domain actually individuates some property of a specific action with respect to the domain concerned. Thus, verbs like 'push,' 'pull,' etc. do not instantiate the result vector which is found in a verb like 'open.' A set of vectors within a given domain form a convex region if and only if the points represented within the vectors can be connected in such a manner that if two distant points x and y are in the domain, any other points between x and y will also be within the same spatial region. This aspect of conceptual spaces implements the sharing of a property (by a number of items) by means of the closeness of points that reflect the same property in a particular domain. The central thesis of the theory of conceptual spaces is that most conceptual spaces for verbs and nouns define convex regions, thereby underlining the point that many exemplars or items can satisfy the same property by staying within the same convex zone. This underscores the prototypical nature of concepts as well, in that prototypical features must be in the central portion of a convex region. Thus, the prototypical features of a chair, for example (four legs and a flat surface with a back for support), must be within the central space of the same convex region. Likewise, for verbs like 'push' or 'pull' the points within the convex region of vectors in the force domain may specify all the prototypical action patterns that designate pulling or pushing.

Needless to say, conceptual spaces reflect the fundamental ingredients of the conceptual fabric that is seamlessly integrated into linguistic structures. This works well especially for linguistic expressions that have a *substantive* (as opposed to notional or abstract) import in visual-spatial domains. The structure of the social, emotional and more abstract domains is not precisely covered under conceptual spaces, which implies that conceptual spaces for verbs such as 'impugn,' 'learn,' 'understand' or for nouns such as 'sincerity,' 'compunction,' 'honesty' may be difficult to characterize, given that the requisite dimensions involved in the respective domains of these words are not immediately clear and compendious. This being so, the structure of the cognitive organization that is encompassed under conceptual spaces is patchy and hence limited. Be that as it may, it is hard to deny that the smaller scale of the cognitive texture captured by conceptual spaces brings out a vaster realm of skeletal frames of our spatial conceptualization of actions, events, states, and properties of entities reflected in language. This may prove useful in view of the embodied nature of scores of linguistically expressed conceptualizations. But this does not, of course, entail that the glimpse into the structure of cognition that we gain by examining linguistic structures evincing conceptual-spatial facets provides evidence that linguistic structures evincing conceptual-spatial facets cannot be shown to have non-spatial cognitive representations. In fact, it is easy to show that many properties of conceptual spaces are such that they may not be fine-grained enough for the individuation of the conceptual substance. Conceptual spaces work best in specifying the geometrical-structural contours of mental constructs that are supposed to be manifest in linguistic structures. What is missing is the actual content that is expressed in linguistic structures and which constitutes the *substance* of conceptual spaces. The geometrical-structural contours of conceptual spaces capture the topological organization of cognitive dimensions or modes of linguistic meanings, while the mental structures that capture the contents of such linguistic elements identify something over above what cognitive dimensions or modes of linguistic meanings express. Hence a cognitive residue is bracketed off, and there is no compelling reason to believe that this cognitive residue is *prima facie* of little consequence.

3.2 Cognition from Language

If the linguistic reflection of mental structures and patterns of our cognitive organization suffices to elevate cognition to a higher level such that language can be viewed as an extension of cognition, linguistic structures in themselves also suffice to lend a very unique character to cognitive processes and mental structures, so much so that cognitive processes and the corresponding mental structures cease to possess the shape they do without language. This issue takes on a completely different character as one takes into consideration the position of Whorf (1956) who adopted the position that it is a specific language that determines a specific way of thinking—a thesis which is known as the *linguistic relativity hypothesis* or the *Sapir-Whorf Hypothesis* (named after the linguists Benjamin Lee Whorf and Edward Sapir). Whorf came to this conclusion after examining and studying the Hopi language and, in particular, the Eskimo language in which different words for different types and shades of snow are found (but see for a different view, Pullum 1991). This motivated him to hypothesize that specific languages humans speak determine the structure of human thoughts as well as the reality our thinking and behavior assume. Here, it must be noted that the original emphasis has been on the linguistic determination of the content or interpretative categories of thought rather than of processes or cognitive mechanisms of thought (see Lucy 1992). There is independent evidence that this thesis can be at least weakly supported. For example, Kay and Kempton (1984) have shown that color perception and categorization, especially the perception of focal colors, can be shaped by the color terms extant in languages, although Berlin and Kay (1969) much earlier attempted to establish that the *universal* cognitive mechanisms of color perception determine the character of (focal) color terms across languages organized in terms of levels since all languages have the same underlying structure of basic color terms. Taken to its extreme, this hypothesis may imply that there would be as many language-respective cognitive patterns as there are languages in the world. That is, around 7000 languages on this planet will give rise to 7000 different interpretative structures of conceptualizing the world. This is not merely absurd but also cognitively opaque. There cannot be any cognitively constrained profiles that can envelop or encompass as many biologically plausible cognitive patterns as there are, had been, and would be languages in the world. It is not clear how one can even include languages that have once existed but are now extinct because the specific cognitive patterns these by-now-extinct languages had instantiated long ago are to be believed to be eliminated. Simply speaking, even if biological variation need not and often does not certainly catch up with this scale of cognitive diversity and variation which appears to be more or less opaque to, or at least not tamed by, the operative biological processes, there would not exist enough cognitive constraints that can filter out possible cognitive profiles or patterns that parallel all natural languages that have existed, presently exist and will emerge in the future.

Interestingly, it would definitely be a modest claim that patterns of cognitive processes may be shaped by the cultures in which humans grow up, and this has indeed been found in certain well-balanced differences in thinking styles (holistic vs. analytical) in Asians and Westerners (see Nisbett 2003). Asians tend to take into account the entire field in their attentional focus when conceptualizing the object concerned, whereas Westerners have been found to focus on the singular items that require analytical organization of the internal parts of the visual scene or object at hand. McWhorter (2014) has of late looked at the whole issue from an entirely different angle. He points out that linguistic relativism is often advanced not merely to show that cognitive profiles determined by different languages differ but also to uplift the socio-cultural status of less recognized, socially suppressed and 'exotic' languages by highlighting the unexplored cognitive profiles of the speakers. However, the crippling problem that surrounds the strong version of linguistic relativism—whether in its cultural or cognitive avatar—is that distinctions that contribute to diversity in cognitive profiles are unmistakably emphasized and so magnified, but the accompanying similarities in cognitive profiles are not concomitantly revealed or foregrounded.

What is at least plausible is that linguistic structures (re)shape or mold certain structural representations of thoughts which would not have taken the form they have without the linguistic molding. Consider, for example, the case of reduplicative nonfinite verbal forms in Bengali.

(26) se khabar **khete** **khete** kotha bolchilo
he food eat-NONF eat-NONF words speak-PAST-IMPERFECT
'He/she was speaking while eating food'.

(27) amra **jitte** **jitte** here gelam
we win-NONF win-NONF losing go-PAST
'We lost coming closer to wining'.

The reduplicative nonfinite verbal forms in (26–27) have been marked in bold. What is striking is that the exact nonfinite verbal forms used here do not mean the same thing when they are used without any reduplication. That is, the verbal form 'khete' means something different from what its reduplicative counterpart does, and the same thing holds true for 'jitte.' Without reduplication, they behave like *to*-infinitives in English. What is of particular concern here is that the nonfinite reduplicative verbal form isolates the imperfective (not completed) temporal portion of an event designated by the verb concerned. That is, this nonfinite reduplicative verbal form signifies the non-completion and durative occurrence of the specific event encoded in the verb. The durative occurrence of the specific event may be iterative or non-iterative. None of the sentences in (26–27) has an iterative reading, but the following sentence seems to have an iterative reading.

(28) Raka chheletake **marte** **marte** niye galo
Raka boy-CLASS beat-NONF beat-NONF taking go-PAST
'Raka took away the boy (repeatedly) beating him'.

[NONF = nonfinite verb form, PAST = past tense, IMPERFECT = imperfective aspect, CLASS = classifier].

The particular conceptualization individuated by this nonfinite reduplicative verbal form is thus akin to the kind of summative processing described in example (20) in Sect. 3.1. What is distinctive here is that the summative view of an event can be underspecified with respect to its iterative dimension such that the event designated takes on a sort of reified character by abstracting away from its real-time occurrence in a given context. It seems as if events which are contingent processes are lifted from their natural contexts of occurrences and then elevated to a level of conceptualization at which events morph into processes arrested and frozen. The specific thought structure that the nonfinite reduplicative verbal form encodes or gives rise to is uniquely specified by such a reduplicative verbal form in Bengali (and other South Asian languages). This is not, of course, available in English. Therefore, here is a case where the structure of thoughts is sculpted from linguistic structures that are in a relevant sense language-respective. The same point can also be raised about the following expressions in English.

(29) What do you wonder who has bought__?
(30) *How do you wonder who went to the party__?

It is important to understand that the sentence in (30) represents a thought which is ineffable at least in English, whereas the sentence in (29) seems to be fine as long as the *Wh*-phrase 'what' is traced back to the object position of the verb 'bought,' as indicated by the dash. Although (30) also involves a *Wh*-phrase and reflects a thought which taken in itself is perfectly legitimate, the sentence is not well-formed as per the syntactic structure of English. Now, the essential point here is that even though there could be nothing about concepts and conceptualizations that prevent a thought from being realized, forms of linguistic expressions may impose their own constraints on such otherwise expressible but not actually expressed thoughts. Thus, thoughts that can be expressed through syntactic forms as can be found in an example like (29) have made it to the mark of being expressible in English, while the thought associated with (30) has not enjoyed the same condition. This indicates that some thoughts are amenable to syntactic expression not because they have something special or hold some privileged status but because some language-respective syntactic constraints on form-meaning matching block some otherwise expressible thoughts from being *syntactically* manifested.

Overall, the discussion just above appears to convey the impression that thoughts are intangible abstract entities or categories that are carried by or contained in syntactic expressions which may or may not vary across languages. This particular idea may also turn on the complex and contentious question of whether or not thoughts are themselves couched in language. Strong positions on this matter have been advanced by various scholars. As a matter of fact, just like the linguistic relativity hypothesis, this issue has got both strong and weak versions depending on how one slices off the part of thought from the very structure of thought that is not considered essentially linguistic. While both Davidson (1984) and Dennett (1991) have vouched for the *ontological* dependence of thought on language, Grice (1989) has advanced the opposite view supporting an *epistemological* priority of thought over language. It is vital to recognize that most people who advocate the (ontological) primacy of language over thought have in mind the propositional structure of thoughts. That is, the belief-like structure of many thoughts (which can be tested by the admissibility of *that*-complements by certain verbs like 'think,' 'believe,' 'feel,' etc.) is what attracts most philosophers of language who tend to think that the structure of thoughts, especially of propositional thoughts, derives from, and for others, is even akin to propositional linguistic expressions (see also Carruthers 1996; Crane 2014). Clearly not all actual thoughts can be rendered in language, for there are visual, tactile, auditory, or even olfactory thoughts. If it were true that all thoughts that are cognitively plausible are identical to linguistic expressions, there would be no room for thoughts that are non-linguistic, and we would be compelled to conclude that all sorts of complex reasoning and patterns of complex representations that have thought-like structures executed in sensory-motor, emotional and other conceptual domains must be linguistic. This is outright absurd. Likewise, one would equally be forced to assert that all linguistic expressions that generally convey thoughts exhaust all possible forms which thoughts can assume. Closer inspection soon reveals that this is also untenable on the grounds that many of our thoughts are repressed thoughts which never reach the level of consciousness that can make them explicit (thoughts in dreams, Freudian thoughts, hallucinatory thoughts, etc.). Moreover, this dogmatic stance on the relation between thought and language would also imply that human beings cease to be thinking beings when they lose language (as in aphasia) or do not acquire any language (cases of infants and feral children). Clearly one cannot afford to adopt this position on reasonable grounds.

What is at least justifiably valid is that linguistic structures can be viewed as streams of thoughts. This view holds especially when one scans linguistic structures as they are used in real time. A close and systematic examination of linguistic structures in use may thus unmask the deployment of thoughts in a sequence that unfolds from thought conglomerates in which some thoughts precede others and the paths these thoughts take depend not just on the speaker's inner conceptualization but also on the behaviors, actions, reactions, thoughts and uttered expressions of the hearer(s) (see Chafe 1998, 2018). Clearly, this is evident in any linguistic discourse. The following examples (taken from Chafe [1998]) are illustrative of this common pattern.

(31) I can't believe…
You know mom was pretty brave.
Did I tell you what Verna told me one time?
…that when she lived there alone she kept old light bulbs up in her bed room and if she heard a noise at night, she would take a light bulb and throw it on the cement walk and it'd pop…just like a pistol going off.

As pointed out by Chafe, the above utterances by some speaker aim to establish an 'epistemological space,' in that the speaker intends to convey her thoughts about some incredible event which the hearer was not apparently aware of earlier and which piqued the curiosity of the hearer as well as of the speaker. The flow of thoughts conveyed through these utterances has a nesting of different levels starting with the expression of incredibility, with the establishment of the point to be made and the description of the source of the information remaining in the middle, and ending with a climax that marks the culmination of the thought

3 Cognition from Language or Language from Cognition?

sequence. It may be noted that the first two lines in (31) signify the expression of incredibility and the third line pinpoints the source of the information, and the rest of the utterance serves to establish and elaborate on the point that is made, finally setting the climax for the hearer who may appreciate the point better now. The entire complex of thoughts is said to emanate from the speaker's 'semi-conscious' state of mind which is taken to be the state of mind that is not fully activated such that the thoughts to be conveyed are not in the pool of given pieces of information, but these thoughts can be accessed and brought into focus for elaboration. Since the speaker of (31) brings into focus the story about the bravery of her mother which required a longer period of time for its full elaboration than is usual for an 'active' state of consciousness, such thoughts are supposed to issue from the semiconscious state of mind. Another point relevant to the discussion is raised by Chafe. He thinks that a singular focus of consciousness in which a portion of the entire thought complex is spelled out must contain the expression of a single 'idea' which would include the articulation of an event or state. The utterance from '…that when she lived there alone' to 'a pistol going off' is not under the ambit of a singular focus of consciousness; rather, the whole utterance is broken into pieces that a sequence of foci of consciousness in the speaker encompass. Thus, a singular focus of consciousness may pack up more than one articulation of an event or state if and only if these events or states are conceptually encapsulated into one single mental event. Needless to say, these foci of consciousness are partly shaped and maneuvered by the assumptions of the speaker about the epistemic status in the hearer's mind as well.

In all, it appears that the flow of thoughts in any linguistic interaction rides on the flow of linguistic expressions that are delivered and understood in suitably organized pieces of structure. Even though this evidently uncovers the *trajectories* of thoughts in action that the human mind projects or tends to form, this is not tantamount to the (re)structuring of cognitive structures or processes by language, and hence, this does not unambiguously show how the underlying thoughts which are exemplary cognitive structures or processes have been (re)structured by linguistic structures. What is sufficiently clear is that the structural constraints implicit in linguistic structures mold the flow of thoughts, but from this, it is hard to arrive at the conclusion that the internal or *intrinsic* structure of thoughts itself derives from that of linguistic expressions. Thus, for instance, if the speaker had stopped by uttering only the fragment 'you know mom was,' the expression would not have conveyed any structured idea or thought. The relevant structural constraint of the linguistic structure is that the predicate that is true of the argument 'mom' must be released and thus expressed. The listener may, of course, fill in the gap in such a case, but the very act of filling in the gap left open by the speaker suggests that thoughts can assume their own form in the absence of any recognizable linguistic expression. That the gap-filling can be constrained by the *structural* contours of linguistic expressions (the predicate 'pretty brave,' e.g., in the case mentioned above) is a different issue altogether.

There is a sense in which the operational character of cognitive processes is *intrinsically* reorganized and restructured by linguistic structures. The best examples of this are provided by Clark (1998). When one mentally rehearses sentences, the mental operations that involve thinking and reasoning are intrinsically restructured by the availability of linguistic structures in the mind. This need not, as Clark thinks, result in a massive reorganization of the whole cognitive architecture as language complements brain's own operations and representations which such operations manipulate. In other words, it seems as if the language-less cognitive architecture develops new ways of exploiting a heterogeneous set of internal and external resources all of which now augment the capacity of the mental processes concerned. Similar things occur when one learns mathematics by memorizing numerical tables for multiplication which cannot be held or preserved in memory *in toto* unless there are linguistic labels that help keep track of mathematical symbols or symbol combinations. Learning a new language or new expressions in a language already acquired often inclines the cognitive system to form new patterns of cognitive structures which were previously unthinkable. The manipulation of linguistic expressions in everyday interactions (spoken or written) also helps transform the shape and space of cognitive operations that deal with symbols. Various acts of planning and online deliberations are boosted and hence enhanced by the rendering of the pertinent plans in linguistic expressions, thereby eliminating beforehand many possible crash-prone paths of learning that the cognitive system could have otherwise taken without language. For instance, when one learns new symbolic systems in fields as varied as mathematics, music, art, culture, architecture, business, or any other academic discipline, language constrains learning in such cases by providing a cognitive space in which the symbolic patterns in the relevant domains of learning can be expressed, allocated, manipulated, and formatted for evaluation, remembering, recall, and overall understanding. Successful learning paths in all such cases are, to a large measure, due to our language capacity. This indicates that the cognitive structures and processes that we are endowed with are partly constituted or shaped by linguistic structures.

It would not be an exaggeration to state that the structure of cognition is, at least in part, akin to the structure of language. If this is the case, one may also wonder if the nature of the whole of non-linguistic cognition is (re)structured by language. Demonstrating that other domains of cognition possess a level of organization that is linguistically formatted is not easy. This may also border on verbalizing the whole of cognition, thereby running the risk of magnifying the role of language beyond a certain limit judged reasonable. We are aware that in the whole animal kingdom humans certainly do not excel in capacities in many non-linguistic domains of cognition such as vision or motor activities, although there is no denying the fact that human non-linguistic cognition restructured by language may also be unique in some way. Hence language seems to be the exception for humans when placed within the broader context of the animal world. This motivates the search for a cognitive domain that is formally structured like language. One of the best candidates for the quest for parallels between language and a domain of cognition is the motor system because both systems appear to share a sequential and hierarchical organization of structural elements. The other commonality is that both capacities are rooted in a common network of brain structures in Broca's area, although this need not *directly* play a causal role in instantiating parallels in the formative elements and their organization in the motor system and language. Let's take one suitable example from Jackendoff

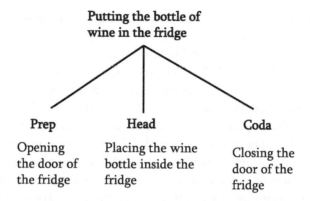

Fig. 3.1 The head vs. non-head organization of a simple motor action

(2007). Putting a bottle of wine in the fridge is a motor activity that can have a dependency-based (head vs. non-head) organization that linguistic structures have. This can be shown in the following way (Fig. 3.1).

Notice that the tree structure of the motor activity for putting a wine bottle inside the fridge does not have a hierarchical organization per se. But it is possible for it to be embedded within another motor action if, for example, the very act of putting a wine bottle inside the fridge is part of another motor activity. Thus, for instance, if the act of putting a wine bottle inside the fridge is itself a preparatory activity for the initiation of the task of adjusting the temperature of the fridge from the digital control display on the exterior of the fridge, then the former will form part of the latter which happens to be the head activity now. In that case, the entire action will look like the following (Fig. 3.2).

It can be observed that each of the constituent actions considered thus far has *itself* a non-hierarchical organization as the Prep, Head and Coda components in Fig. 3.1, for example, branch out from the same root and are thus at the same horizontal level, although the head in Fig. 3.2 has its own branches corresponding to the complexity of the action of adjusting the temperature of the fridge. In fact, it is also possible to recast any of the basic motor actions discussed above into a format that parallels the structure of a phrase within the X-bar schema which is oriented around the structural organization of heads and non-heads in terms of an implicit hierarchy (see for details, Knott 2012).

3 Cognition from Language or Language from Cognition?

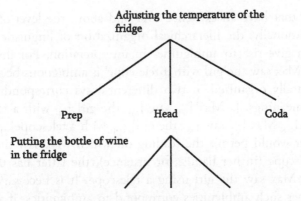

Fig. 3.2 One motor action embedded within another motor action

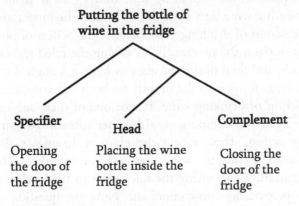

Fig. 3.3 The X-bar organization of a motor action

In that case, the task of putting a bottle of wine in the fridge may also look like what is shown in Fig. 3.3.

It is vital to recognize that even though motor actions can be characterized by having a hierarchical organization, it is not the case that the hierarchical organization of basic motor functions/actions can allow for an unbounded number of levels of (self-)embedding even *in principle*. This is because all motor actions are real-time actions—they are executed in real time. Unlike language, the motor system cannot be viewed as an abstract system that permits potentially infinite levels of (self-)embedding of motor actions. The intrinsic character of the motor system is such that its outputs cannot be considered to be abstract structures *in themselves* that have an existence over and above the level of instantiation. Additionally, the hierarchical organization of linguistic structures may often give rise to ambiguities in interpretation. For instance, the sentence 'Max saw the girl with a telescope' is ambiguous because it can be structurally organized in two different ways corresponding to two different meanings: [$_S$ Max [$_{VP}$ saw [$_{NP}$ the girl [$_{PP}$ with a telescope]$_{NP}$]$_{VP}$]$_S$] and [$_S$ Max [$_{VP}$ saw [$_{NP}$ the girl] [$_{PP}$ with a telescope]$_{VP}$]$_S$]. While the former would permit the reading on which Max saw the girl who had a telescope (in her hands, for instance), the latter has the reading on which Max saw the girl using a telescope. It is necessary to figure out whether such ambiguities correspond to ambiguities, if any, which are to be found in the hierarchical organization of motor actions. Take, for example, the action of pouring wine into a glass to drink wine. The action of pouring wine into a glass is, of course, the preparatory component of the action of drinking wine because the action of pouring wine into a glass initiates the process. Now, raising the filled glass to the level of the mouth, and then tilting the glass by having it aligned with the lips so that the wine flows into the mouth are both sub-components within the head action of drinking wine. Hence one of these sub-components can be the head sub-component with other sub-component being the preparatory action. Thus, tilting the glass by having it aligned with the lips, thereby letting the wine flow into the mouth can be the sub-component head, while raising the filled glass to the level of the mouth can be the preparatory sub-component. Now, the question is: Can the action of raising the filled glass to the level of the mouth be organized *either* as a preparatory sub-component with respect to the head action of tilting the glass *or* as a preparatory component with respect to main higher-level action of drinking wine? There is no doubt about the former case. But in the latter case, one may argue that this entirely depends on our construal of actions which more often than not turns on the linguistically structured conceptualization of action events.

One may thus state that there is no independently recognizable action of drinking wine which is *intrinsically* anchored in the motor system, since drinking wine is a linguistically based conceptualization that is partly cultural. Hence, as the argument may go, what are internal to the motor system are the actions of pouring wine into the glass,

3 Cognition from Language or Language from Cognition?

raising the filled glass to the level of the mouth and tilting the glass. Many motor actions that we linguistically conceptualize and talk about can thus be recast in a different way. They may be viewed as redescribed and reformatted linguistic conceptualizations of motor actions. If this is the case, one would be hard-pressed to conclude that the internal organization of motor actions is modularly interpreted. Rather, what is clear is that the structural organization of many motor actions as conceptualized by humans is a reflex of the conceptual system of categorizations that language constitutes and gives rise to. If the conceptual texture of categorization is largely structured by language in a way that we barely comprehend when engaging in motor activities, we may do better by examining its properties so as to determine how the fabric of cognition is molded by language.

Categorization and perception are domains of cognitive capacities that quite conspicuously and vividly reflect the imprints of linguistic structuring. The linguistic structuring of cognition can be telescoped into the form of integration and coupling of categorization and perception in order that the structure of categorization reflects linguistically structured perception and that the structure of perception reflects linguistically structured categorization. The world in its whole may be an unstructured or unorganized collection of a variety of distinct kinds of entities and processes. But when linguistic labels (words, for example) or linguistic structures portion out different groups of entities and processes, these groups take on a new form for the mind. Things that have remained thus far unfamiliar and strange cease to be unknown and thus become more common and ordinary once they are customized by linguistic structures. This is because language may be supposed to reflect the ontological structure of the world, regardless of whether the reflection preserves any kind of isomorphism or not. Thus, it seems as if the world is disclosed to us in a completely new and reshaped fashion when language captures and (re)configures the structure of the world. Language learning is, in a certain sense, tantamount to learning how to categorize the world as our conceptual world is shaped by language in ways that carry marks of linguistic demarcation. For example, the difference between 'lion,' 'the lion' and 'lions' can be cashed out in terms of the distinctions in categories picked up by these noun phrases. The difference between 'lion,' 'the lion' and 'lions' is definitely a *grammaticalized* difference—a difference that is linguistically shaped but is not intrinsic to any other cognitive system. But this difference turns out to have cognitive ramifications, in that the grammaticalized distinctions between 'lion,' 'the lion' and 'lions' correspond to distinctions in categories pertaining to the animal 'lion.' The word 'lion' picks up an indivisible mass where the recognizable body parts of lions do not matter. Hence, it makes more sense to say 'Those soldiers used to eat lion (as a meat) during war.' But the noun phrase 'the lion' may individuate a particular lion or the whole group of lions as animals in a generic sense. In such a case, the entity is not thought of as a whole undivided mass; rather, instances or tokens of lions and the type these tokens exemplify come to be more significant, and hence, we can say 'The lion is an animal of immense strength and power' or 'The lion came to the southern part of the jungle yesterday.' Likewise, 'lions' as a noun phrase isolates a group of animals having a distinct biological category, and thus, this form has a generic meaning of the animal called 'lion' which is clearly manifest in a sentence like 'Lions are way stronger than hyenas.'

Notice that each particular grammaticalized form of the noun phrases isolates a different way of categorizing the same entity. The same entity can thus be categorized either as an undifferentiated mass or as an entity that shifts between a token realization and a type realization, or even as a generic group of distinguishable tokens. This is also clearly evident in grammaticalized forms of aspect marked in verb phrases. The following examples illustrate this well.

(32) Glynn knows (*is knowing) the difference between cowardice and valor.
(33) She was walking across the garden *in/for hours.
(34) The engineers built the architecture in/for a year.
(35) All the workers together pushed for/*in an hour the car that got stuck in the mud.
(36) The poor man died instantly/*? in/*for an hour.

What these examples show is that events can also be categorized in different ways that neatly correspond to grammaticalized forms of verbal

3 Cognition from Language or Language from Cognition?   129

aspect. The example (32) denotes an event which has the temporal organization of states which remain invariant for all times—which explains why a progressive form of the verb 'know' is invalid in English. The example in (33) designates an activity which is unbounded in space and time in the sense that its termination is unknown or indeterminate, and hence, the preposition 'in' that marks termination (also called 'telicity') is at odds with the very character of the event in question. The examples in (34) and (35) both individuate an activity which has a definite end point but whose process has an unknown duration. That is the reason why such cases can be construed in two different ways that hinge on the question of whether or not the process component in the event conceptualized is terminated. While (34) may be supposed to be construed in both ways, (35) appears to have only one interpretation on which it is not obvious that the process component in the event of pushing the car terminated and thus it remains unknown. On the other hand, (36) characterizes an event that does not have any process component such that it seems as if it occurs in no time. What is at least clear is that if an event is linguistically conceptualized in such a way that it allows for room for a process, however long or short it is, the corresponding linguistic structure that contains a verb designating the event in particular can permit progressive forms of aspect in the clause. The following examples are relevant to the point made here.

(37) The boy ran *for/in an hour to the shop located near the workshop.
(38) The boy ran for/*in an hour towards the shop located near the workshop

Thus, the same event can be categorized in two different ways depending on how the event is delimited by some linguistic expression that delimits the process component. The expression 'to the shop ...' in (37) delimits the event of running in such a manner that the event concerned is understood to have terminated, whereas 'towards the shop ...' in (38) does not delimit the activity of running in that way, thereby allowing the event to terminate at an undefined point in time. Verbs that designate events which are usually thought to have no process component (such as 'die,' 'arrive,' 'win') can, in fact, be construed in a manner that reveals the flexibility in linguistic categorization. The examples (39–40) show this quite evidently.

(39) The girl's mom would be arriving soon.
(40) When one is winning, nothing hurts.

Therefore, it is not just entities that can be categorized in different ways—processes that are by their very nature dynamic can also be categorized in heterogeneous ways that are governed by grammatically distinguished forms. There is no doubt that the ability to categorize entities and processes at various scales of abstraction confers on our cognitive system a high degree of detachment from the outside world, primarily because one can conceptually organize events and/or entities a multi-dimensional space of possibilities by locating them at points that are not instantiated in reality and perhaps never will. This can be influenced, rather than determined, by the number of conceptual distinctions available in a particular language. Thus, a state that cannot be construed in the progressive aspectual form in English ('know,' 'understand,' 'love,' 'like,' etc.) can be construed in that way in another language. The examples (41) and (42) from Bengali and Turkish, respectively, show this quite clearly.

(41) se   ajkal   Roy-er   kachh theke sabkichhu **jan**-chhe
     he/she nowadays   Roy-GEN   near from everything know-PROG
     'He/she nowadays gets to know everything from Roy'.
(42) Hasan fazla cabuk   konus-tug-un-u   **bil**-iyor-du
     Hasan too   fast   talk-NOML-3SG-ACC   know-PROG-PAST
     'Hasan knew that he was speaking too fast'. (Kornfilt 1997)

(GEN = genitive case, PROG = progressive aspectual marker, NOML = nominalization, ACC = accusative case, SG = singular number, PAST = past tense)

Even in English, it is sometimes possible to have such verbs in the progressive form, especially when the dynamic ongoing portion of an event is foregrounded in the relevant categorization of events. Thus, the following cases are perfectly acceptable.

(43) I have been **meaning** to ask you out for dinner tonight.
(44) Of course, I was **forgetting** something. It is Steve's birthday.

The verbs that designate states have been marked in bold in (41–44). What turns out to be significant here is that linguistic categorization is not always held hostage to the dictates of language-specific constraints on form-meaning mappings. Rather, the conceptual system of categorizations more often than not flows freely through language. The nature of linguistically formatted categorization is also manifest in the use of adjectives in language. It may be observed that a distinction is often made between the attributive use of adjectives and its predicative use. It should be noted that this distinction, which is otherwise grounded in the clausal structure of a sentence, can also be cashed out in terms of cognitive contributions to categorization. Let's look at simpler cases to understand the point at stake.

(45) Jeet is a good man.
(46) Jeet is good.

Notice that the attributive use of the adjective 'good' is localized in the domain of the noun phrase 'a ... man,' while the predicative use is anchored in the domain of the verb phrase 'is good.' Now, the question is: Do (45) and (46) convey the same meaning or carry the same content by virtue of the fact that both sentences contain the same adjective 'good'? Clearly there seems to be a difference in meanings between (45) and (46), in that the sentence (45) talks about the person 'Jeet,' who is then characterized as a good man, but (46) conveys the impression that Jeet has a certain quality in his personality by virtue of which he is considered good. This distinction is also conspicuously present in the logical representations of the sentences, as shown below in (45') and (46'), respectively.

(45') $\lambda x [man (x) \wedge good(x)]$ (Jeet)
(46') $good(j)$ $\qquad$ [j = Jeet]

What (45') in essence says is that Jeet belongs to the set of men *and* the set of individuals who are good (this becomes possible after 'Jeet' takes the value of x upon lambda-abstraction), whereas (46') tells us that Jeet belongs to the set of good individuals. That is, as per (45') Jeet possesses the property of being a man and good, and (46') tells us that Jeet has the property of being good, which is, after all, a first-order property. What is evident above is that the attributive use of adjectives straddles two categories or classes, while the predicative use of adjectives involves only one category. This indicates that the attribute use of adjectives carries a greater amount of categorization complexity than the predicative use of adjectives—something that is made possible via language. Importantly, these categories bear different characteristics that are ultimately related to the way we perceive the world and the perceived world is conceptualized by us. The distinction in categories that is pertinent in this context is the distinction between *individual-level* (adjective) predicates and *stage-level* (adjective) predicates (Carlson 1977)—the former refers to the permanent and salient property of individuals or entities, and the latter to the more transient and momentary or extrinsic property of individuals or entities. The examples below are indicative of the distinction in question vis-à-vis the attributive and predicative uses of adjectives.

(47) Daphne still remembers her fond teacher.
(48) *Daphne still remembers her teacher who is/was fond.
(49) Food items are now available for the young boys.
(50) The available food items are for the young boys.
(51) He made occasional visits to the dilapidated house.
(52) *He made visits to the dilapidated house which are/were occasional.
(53) We hope the kids are awake now.
(54) *We hope the awake kids are in the kitchen.

It is clear from the examples above that any transition in the case of certain adjectives from the predicative use of adjectives to the attributive use or vice versa leads to either ungrammaticality or a change in their individual-level/stage-level status. Note that 'fond' as an adjective is

3 Cognition from Language or Language from Cognition?

generally used predicatively meaning 'affectionate towards,' and hence, it requires two arguments (the person of whom one is fond and the person who is fond of someone). However, when it is used attributively, it retains the meaning of something or someone being affectionate, but it cannot be used predicatively without the preposition 'of' and with the same attributive meaning that categorizes someone or something as loving or affectionate. Clearly 'fond' is an individual-level (adjective) predicate. It seems that even if a transition from a categorially simpler use (i.e., from the predicative use) of 'fond' to a categorially complex use of it (i.e., to the attributive use) is valid within the same domain of the individual-level semantic configuration, a transition in the reverse direction from the categorially complex use to the categorially simpler use of 'fond' is banned. This describes the examples in (47–48). The case in (49–50) is slightly different in the sense that the adjective 'available' in (49) is a stage-level predicate since availability does not designate any permanent indispensable property of food items. Its categorially simpler use makes a smoother transition to a categorially complex use in (50). But what is interesting here is that 'available' in the noun phrase 'the available food items' ceases to be a stage-level predicate since the adjective does not isolate any situational or extrinsic property of food items; rather, it characterizes the class of food items that are so classified by virtue of being available. This is evident enough in the simultaneous use of the same adjective in both pre-nominal and post-nominal positions, as in 'the available food items available' in which the latter is clearly stage level, but the former is not. The examples (51–52) parallel those in (47–48), with 'occasional' being a stage-level (adjective) predicate. Significantly, the case in (53–54) presents a scenario which is opposite to what we have found either in (47–48) or in (51–52), in that a transition from a categorially simpler use of the adjective 'awake' to a categorially complex use is blocked.

In a nutshell, bidirectional transitions between a categorially simpler use of an adjective and its categorially complex use are not *always* allowed within a language. Hence adjectives like 'good' tell us a story which remarkably differs from the one that the adjectives in (47–54) present us. Interestingly, it is crucial to recognize that this issue connects categorization to perception seamlessly. Notice that the examples in

(47–54) unequivocally convey the impression that bidirectional transitions between a categorially simpler use of an adjective and its categorially complex use *may* trigger either ungrammaticality or an alteration in the individual-level/stage-level status of (adjective) predicates. This implies that an index (binary or otherwise) of categorial complexity in adjectives can signal the validity/invalidity of their perceptual properties to be cashed out in terms of the individual-level vs. stage-level distinctions. In the case of adjectives like 'good,' bidirectional transitions preserve the validity of the perceptual configuration projected by the adjectives concerned, whereas other adjectives (as in (47–54)) indicate the invalidity of their *initial* perceptual character (by means of ungrammaticality or through an alteration in the individual-level/stage-level status) when bidirectional transitions are introduced. This may be traced to the very nature of categorially motivated perception that is encoded in language. Although it is quite plausible that our perceptual organization of the world is non-discrete and continuous with the sensory-motor fields of perception, it is language that imposes its own form of discrete categorial organization on whatever is perceptually transmitted and thereby configured. By making an individual-level vs. stage-level distinction in adjectives, language imposes its own order on the perceptual configurations derived from the sensory-motor fields of perception, and this very distinction interfaces with and attaches to an intrinsically linguistic level of conceptual organization which is more fundamental to both perception and categorization in humans.

3.3 Language–Cognition Synergy

We have observed that if there are ways in which language can be said to derive from cognition, there are also other equally convincing ways in which cognition can be reckoned to be rooted in language. It would not be unreasonable to concede that one cannot faithfully elevate the logical directionality from language to cognition over the logical directionality from cognition to language, or vice versa. Often accounts of the language–cognition interface are advanced in order to support only a single type of logical directionality in language–cognition relations. This is especially done to underscore the primacy of language in the fabric of cognition, or to highlight the grounding role of the underlying cognitive substrate for language. Although arguments from both sides seem equally compelling or at least largely parallel to one another, there is no denying that it is also tempting to side with a particular position with respect to language–cognition relations. Support for the primacy of language in the (re)structuring of cognition may be taken to provide scaffolding for the idea that the whole of cognition is essentially linguistic and symbolic. On the other hand, attempts to establish that language is itself derived from the basic format of cognition may be viewed as ways of attenuating claims about the ontological uniqueness and primacy of language with respect to cognition. What this chapter has shown is that no matter how cogent the arguments for either of the views are, there is reason to believe that the question of whether language originates from cognition or cognition emanates from language cannot be resolved, crucially because the question itself revolves around language and is curiously rendered in language.

One cannot get outside of language and then pose the question of ontological primacy between language and cognition so that it could be resolved once and for all. The matter is much more delicate and subtle than it appears to be. It is important to realize that if one claims that language derives from cognition, one has to demonstrate in unequivocal terms that all properties and aspects of language can be smoothly translated into the properties and aspects of our cognitive organization in a such a way that all properties that are intrinsically linguistic will remain preserved over the translation from the system of language into the system of cognitive organization and no further translation in the reverse direction would be possible. Likewise, if one aims to show that the cognitive system itself derives from language, it has to be unquestionably demonstrated that cognitive structures and processes are constituted only and entirely by linguistic structures and processes that language involves. Note that the first case starts with the individuation of the properties and features of language and in the second case cognitive properties are finally traced to the properties of the linguistic system. Thus, in both cases the individuation of the properties and features of language relevant to cognition assumes utmost significance, and if this is the case, one never gets out of the condition in which the individuation of the properties of language relevant to cognition can be bracketed off. Hence we never end up moving out of the chains of circularity in which language–cognition relations are couched in terms of a certain kind of directionality.

We may thus recognize that instead of trying to find out what comes from what, we can consider the system of language itself to be a cognitive system and, in doing so, endow language with a status that makes it an *intrinsic* part of the cognitive architecture itself. It is with this belief that we feel justified in probing into the question of how language can provide entry into the domain of cognition. This is so because if something is part of some larger whole, the structure and organization of the part itself can provide deeper insights into the structure of the whole just as the structure of the whole can tell us a lot about its parts. But when very little is understood about the whole, the parts can themselves prove to be extremely valuable and illuminating for one's understanding of the whole. Here a similar stance needs to be adopted. It seems reasonable to maintain that neither language nor cognition holds any supremacy or primacy over the other. Even though we deem language to be an intrinsic part of our cognitive organization, it would be an understatement to say that language is *just* a part of our cognitive organization. Rather, language can be said to be the central cognitive system which is a *sui generis* part of the cognitive system. This special part–whole relation has grown out of the synergy that made it possible for language and cognition to bond with one another in the first place. There may have been some point in time in the evolutionary past when our hominid ancestors were without language or any other mode of symbolic communication. So cognition at that stage was certainly non-linguistic. But when language invaded the cognitive territory and then became fully integrated into the cognitive infrastructure, it is not merely language that was shaped by the cognitive substrate—the cognitive substrate too perhaps gradually became embedded into the overall organization of the mind that language gave rise to, thereby paving the way for the meshing of cognition with language. This could have forged an inalienable coupling between language and cognition hardening into a kind of synergy that made it well-nigh impossible to pull language apart from cognition and possibly vice versa.

It is this synergistic relationship between language and cognition that we appeal to when seeking to understand the formal structure of cognition by looking into the formal properties of language. Biological relations of instantiation and reduction are unnecessary in understanding this relationship since the synergy itself goes beyond matters of instantiation and reduction within the biological substrate. The biological substrate may have provided the platform for the synergy to get off the ground, but the actual nature and form of the synergy *in itself* is outside of biological relations of instantiation and reduction which are oriented around cross-scale translations. Thus, we think it is wise to explore, not of course in a derivative manner, the structure of language itself to understand more about the structure of cognition, especially when the aim is to bypass relations of biological grounding.

3.4 Summary

By exploring language–cognition relations in various forms in different contexts of their interactions, this chapter has attempted to show that it is perhaps futile to seek to derive language from cognition, or to extract cognition from the system of language *only* in a unidirectional fashion. As the chapter has argued, it is much more interesting to accept that the relation of derivation between language and cognition can go either way because doing so permits one to view either side of the relation most succinctly. Equipped with such a two-sided view of the broader scenario, we can undertake to scrutinize properties of language in order to know more about cognition or even vice versa, without of course presupposing anything about derivation relations between the two. Each such goal will undeniably carry its own perspectival requirements and other concomitant conditions, but this will certainly be better served by the background understanding that the investigation concerned assumes no ontological primacy of one system over the other.

References

Berlin, B., & Kay, P. (1969). *Basic Color Terms: Their Universality and Evolution*. Berkeley: University of California Press.
Bickerton, D. (2014). *More Than Nature Needs*. Cambridge, MA: Harvard University Press.
Carlson, G. (1977). *Reference to Kinds in English*. Doctoral dissertation, University of Massachusetts.
Carruthers, P. (1996). *Language, Thought and Consciousness: An Essay in Philosophical Psychology*. Cambridge: Cambridge University Press.
Chafe, W. (1998). Language and the flow of thought. In M. Tomasello (Ed.), *The New Psychology of Language* (pp. 68–93). New Jersey: Lawrence Erlbaum.
Chafe, W. (2018). *Thought-Based Linguistics: How Languages Turn Thoughts into Sounds*. Cambridge: Cambridge University Press.
Clark, A. (1997). *Being There: Putting Brain, Body and World Together*. Cambridge: MIT Press.
Clark, A. (1998). Magic words: How language augments human computation. In P. Carruthers & J. Boucher (Eds.), *Language and Thought: Interdisciplinary Themes* (pp. 162–183). Cambridge: Cambridge University Press.
Clark, A. (2008). *Supersizing the Mind: Embodiment, Action, and Cognitive Extension*. New York: Oxford University Press.
Crane, T. (2014). *Aspects of Psychologism*. Cambridge, MA: Harvard University Press.
Davidson, D. (1984). *Inquiries into Truth and Interpretation*. Oxford: Clarendon Press.
Dennett, D. (1991). *Consciousness Explained*. New York: Little Brown.
de Villiers, J. (2007). The interface of language and theory of mind. *Lingua, 117*(11), 1858–1878.
Gärdenfors, P. (2004). *Conceptual Spaces: The Geometry of Thought*. Cambridge: MIT Press.
Gärdenfors, P. (2014). *The Geometry of Meaning: Semantics Based on Conceptual Spaces*. Cambridge: MIT Press.
Gopnik, A., & Astington, J. W. (2000). Theory of mind. In K. Lee (Ed.), *Childhood Cognitive Development: The Essential Readings* (pp. 175–200). Oxford: Blackwell.

Grice, H. P. (1989). *Studies in the Way of Words.* Cambridge, MA: Harvard University Press.
Hirstein, W. (2005). *Brain Fiction: Self Deception and the Riddle of Confabulation.* Cambridge: MIT Press.
Hurford, J. R. (2007). *The Origins of Meaning: Language in the Light of Evolution.* New York: Oxford University Press.
Jackendoff, R. (1983). *Semantics and Cognition.* Cambridge: MIT Press.
Jackendoff, R. (2002). *Foundations of Language: Brain, Meaning, Grammar, Evolution.* New York: Oxford University Press.
Jackendoff, R. (2007). *Language, Culture, Consciousness: Essays on Mental Structure.* Cambridge: MIT Press.
Johnson, M. (2018). *The Aesthetics of Meaning and Thought—The Bodily Roots of Philosophy, Science, Morality, and Art.* Chicago: Chicago University Press.
Kay, P., & Kempton, W. (1984). What is Sapir-Whorf hypothesis? *American Anthropologist, 86,* 65–79.
Knott, A. (2012). *Sensorimotor Cognition and Natural Language Syntax.* Cambridge: MIT Press.
Köhler, W. (1927). *The Mentality of Apes.* London: Routledge & Kegan Paul.
Kornfilt, J. (1997). *Turkish Grammar.* London: Routledge.
Lakoff, G. (1987). *Women, Fire and Dangerous Things.* Chicago: Chicago University Press.
Langacker, R. (1987). *Foundations of Cognitive Grammar.* Stanford: Stanford University Press.
Larson, R. (1989). *Light Predicate Raising* (MIT Lexicon Project Working Papers No. 27).
Lucy, J. A. (1992). *Language Diversity and Thought: A Reformulation of the Linguistic Relativity Hypothesis.* Cambridge: Cambridge University Press.
McWhorter, J. (2014). *The Language Hoax: Why the World Looks the Same in Any Language.* New York: Oxford University Press.
Millikan, R. G. (2005). *Language: A Biological Model.* New York: Oxford University Press.
Millikan, R. G. (2018). *Beyond Concepts: Unicepts, Language, and Natural Information.* New York: Oxford University Press.
Nisbett, R. E. (2003). *The Geography of Thought.* New York: The Free Press.
Pullum, G. K. (1991). *The Great Eskimo Vocabulary Hoax.* Chicago: Chicago University Press.
Talmy, L. (2000). *Towards a Cognitive Semantics* (2 vols.). Cambridge: MIT Press.

Tomasello, M., & Call, J. (1997). *Primate Cognition*. New York: Oxford University Press.
Tomasello, M. (1999). *The Cultural Origins of Human Cognition*. Cambridge, MA: Harvard University Press.
Tomasello, M. (2019). *Becoming Human: A Theory of Ontogeny*. Cambridge, MA: Harvard University Press.
Vygotsky, L. S. (1962). *Thought and Language*. Cambridge: MIT Press.
Whorf, B. L., & Carroll, J. (Eds.). (1956). *Language, Thought and Reality*. Cambridge: MIT Press.

4

Linguistic Structures as Cognitive Structures

The previous chapter has delineated the ways in which language can be derived from the fundamental organization of cognition and the manifold character of cognition can be distilled from the structural details of language. We have observed that the situation is not as neat and clean as many people may seem to believe. For every cogent argument for the ontological primacy of cognition over language, we can have an equally cogent argument that supports the ontological primacy of language over cognition. Notwithstanding this parallel in the ontologically grounded relations between language and cognition, there seems to be some residue of cognition that is not exhaustively covered by language–cognition relations. This is because cognition as a whole is more than what linguistic cognition constitutes and thus encompasses. Besides, making language the fulcrum of language–cognition relations serves our purpose better, in that the immediate amenability of language to an investigation into its structural properties makes it more than suitable for penetration into the realm of cognition. On the other hand, if we reckon cognition itself to be the fulcrum of language–cognition relations, many properties of cognition cannot be tapped into as we would be trying to understand something through itself, thereby inviting the ineluctable problem of circularity. Further, if we want to penetrate into the realm of cognition by bypassing biological relations of derivation and instantiation, a reasonable approach would be to employ a strategy that consists in the deployment of an intermediate tool that can comfortably gain entry into what we want to understand. The structure of (linguistic) cognition being what we intend to understand and the relation to the underlying biological infrastructure not being the intermediate tool that we want to deploy, language seems to be the most optimal bridge that can take us inside the arena of cognition. Here, cognition cannot in itself be the bridge precisely because (linguistic) cognition cannot help us understand the structure of cognition unless we understand and analyze language to have a better grasp of its cognitive basis. An analogy may be helpful here. Suppose by looking into the organization of the perceptual system which evinces various aspects of cognition, we want to understand the structure of cognition in a way that does not involve trespassing on the biological domain in order to accomplish this goal. In this case, we may do better by probing into the structural and representational properties of the perceptual system itself. We would like to achieve the same in the current case.

4.1 The Cognitive Constitution of Linguistic Structures

We are now geared up to explore the intrinsic cognitive structuring of linguistic structures. To this end, we need to examine a heterogeneous range of cases of linguistic phenomena that naturally admit of an explanation by means of which linguistic structures can be viewed as cognitive structures. There are numerous linguistic structures that appear to be exclusively *linguistic* in displaying no noticeable cognitive facets, but they will be shown to be cognitively constituted. Thus, the following subsections will consider a number of cases of linguistic phenomena such as variable binding, quantification, complex predicates, and word order that are logically intricate and defy a unifying linguistic explanation. Then, the overarching aim is to demonstrate that the structural formats of many linguistic phenomena unmask a hidden cognitive texture which has been hitherto overlooked and unexplored. Many of the principles underpinning the cognitive constitution of linguistic structures may also be taken to be the 'laws' of our underlying cognitive organization.

4.1.1 Variable Binding

Let's now look at the following sentences.

(55) The guy over there knows what he wants to do here.
(56) While the baby was asleep, she never rolled over.
(57) Because he never came late, Roy finished off all that was to be done.
(58) She felt tremors on the porch, while Jane was knitting a sweater.
(59) The actor who every woman seeks to meet seems to think she looks ravishing.
(60) As our friend swims in the river that is situated alongside the land cultivated by all farmers, they gear up to look after all of us.

The sentences in (55–60) show what is called *variable binding* which consists in the assignment of a meaning to a pronominal item on the basis of the meaning of a quantifying or non-quantifying item which controls the pronominal item. The consequence of variable binding is that the pronominal item that is controlled by the binding item varies in meaning with that of the binding item. Thus, in (55), for example, the binding item 'the guy over there' controls and hence binds the pronoun 'he,' although it is not, of course, necessary that this binding has to obtain because 'he' *can also* refer to someone other than what 'the guy over there' refers to. Similar considerations apply to (56) too. From this, it seems that any binding item that precedes the variable item bound can control the meaning interpretation of the bound item. The case in (57) is quite interesting since the pronoun 'he' precedes the binding item 'Roy' and can still be bound by 'Roy,' although it is evident that 'he' can also refer to someone else as well. Thus, it looks as if the precedence of the binding item over the item that is bound is not the sole criterion for variable binding. The sentence (58) provides clear evidence that variable binding is sensitive to the precedence relations of the binder and the item bound as the precedence of the pronoun 'she' does not get it bound by 'Jane.' But the examples in (59–60) demonstrate that a mere precedence of the potential binder over the item to be bound is not sufficient to guarantee that variable binding will obtain. Just as 'every woman' does not bind 'she,' the phrase 'all farmers' does not bind 'they.'

It is quite evident that variable binding is a complex phenomenon straddling syntax and semantics. But the amount of variability displayed in variable binding as shown in the chosen fragment of English seems to defy any unified account. What is more important here is the question of whether this variability in variable binding can be couched in cognitive terms so that the patterns observed turn out to have a cognitive import. But before we do that, we need to look at the linguistic analysis of variable binding. It is significant to note that even though a mere surface precedence of the binding item over the item bound does not guarantee variable binding, the hierarchical height relations of the binder and the variable that is bound by the binder matter a lot. This seems to explain why variable binding occurs in (55–56) but not in (59–60), simply because the potential (quantificational) binders 'every woman' and 'all farmers' in (59) and (60), respectively, are deeply embedded within the noun phrases 'the actor who every woman seeks to meet' and 'the land cultivated by all farmers,' respectively. This increases the *depth* of the path that can connect the binder and the variable. The following schematic representations show this clearly.

The dashed arrows in Figs. 4.1 and 4.2 show the path between the binder and the pronominal variable that can connect them. Clearly, the

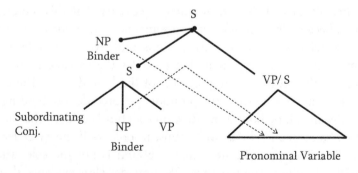

Fig. 4.1 The path between the binder and the variable in (55–56)

4 Linguistic Structures as Cognitive Structures

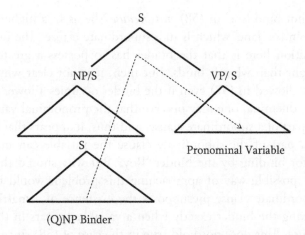

Fig. 4.2 The path between the binder and the variable in (59–60) (NP=noun phrase, QNP=quantificational noun phrase, VP=verb phrase, S=sentence)

path connecting the binder and the variable in Fig. 4.1 is of a lower depth than the one in Fig. 4.2. But note that the (Q)NP binder cannot connect to the pronominal variable in a linear path, and hence, no horizontal path can connect the two. Figure 4.1 represents two possibilities through the two edges projected from S as two different possible trajectories for an NP (as in 55) and an S (as in 56) because (55) has an NP binder and (56) has a subordinating clause containing the NP binder. In short, the paths that can connect the binder and the variable in (55–56) are of a lower depth because the binder has to go *at most* two edges up to be able to dominate the variable from the root node (i.e., S at the top). Thus, the NP binder (i.e., the NP 'the guy over there') in (55) goes one edge up to reach the root S, and the NP binder 'the baby' in (56) goes two edges up to reach the root S at the top. On the other hand, in both (59) and (60) the binder has to move more than two edges up to reach the top S node.

However, the whole account appears to crumble when we look at (57–58). While the binder 'Roy' in the matrix clause can bind the pronoun 'he' in the subordinate clause even though the pronoun in the subordinate clause is in a higher position than the binder, the binder

'Jane' cannot bind 'she' in (58) *just because* 'she' is in a higher position than the binder 'Jane' which is in a subordinate clause. The underlying presupposition here is that the binder has to possess a greater hierarchical height than what it binds. But then, it is not clear why variable binding is allowed in (57) even if the binder occupies a lower position. Given this dilemma, one may observe that the pronominal variable 'he' in (57) is part of a subordinate clause, and thus, it appears that by virtue of being a part of the subordinate clause the variable can make itself available for binding by the binder 'Roy.' But why should this be the case? One possible way of approaching this problem would be to say that a subordinate clause presupposes the existence of a matrix clause, thereby fixing the binder exactly when a variable appears in the subordinate clause. This does not hold true in the case of (58) since a matrix clause does not presuppose the existence of a subordinate clause in the way a subordinate clause presupposes the existence of a matrix clause. In other words, the matrix clause 'she felt tremors on the porch' in (58) can stand alone when it appears at the front, and this frees the variable 'she' from a possible binding by any binder that may appear in the subordinate clause. Even so, we are left without any account of the (hierarchical) height relations between the binder and the variable.

What seems at least clear is that if the binder is hierarchically higher than what it binds, the variable would be bound by the binder. But this does not work in a converse fashion as well. That is, if the variable is at a higher position than the binder, it does not necessarily follow that the variable cannot in any case be bound by the binder, as observed in (57). Thus, we are again in a logjam. Does this ultimately mean that hierarchical ordering is not really adequate in explicating the patterns of variable binding that we find within natural language? Should we then eschew hierarchical ordering in order to appeal to some other aspect of the structural organization of language with a view to accounting for the nature of variable binding? For all we know, it is not fully clear that hierarchical ordering is entirely irrelevant to the nature of variable binding, as observed above. That is why we find the following patterns.

(61) The university followed its guidelines on the admission procedure.
(62) Its guidelines on the admission procedure mattered to the university.

While the noun phrase 'the university' binds 'its' in (61) on its most natural interpretation, it is not immediately obvious that 'its' in (62) is automatically bound by 'the university.' It is evident that this difference in variable binding is hard to explain without making reference to the fact that varying hierarchical orderings of binders and the variables they bind make interpretations vary across sentences. Hence, it is perhaps safer in a pertinent sense to assume that hierarchical relations of binders and variables are part, but not the whole, of the story that variable binding gives rise to. This holds even though the underlying logic of variable binding is fully transparent to, or simply independent of, any hierarchical relations between the binder and what is bound. This is because in terms of the logical analysis binding consists in binding the variable by means of an operator (say, a lambda operator) which also binds another variable created as a placeholder for the binder.[1] This ensures that what is bound and the placeholder variable for the binder are under the scope of the same operator (see Reuland 2011). But this leaves more questions unanswered rather than answered.

At this juncture, we may attempt to figure out how variable binding in natural language is cognitively constituted and not merely influenced or governed by facets of our cognitive organization. In order to do this, we have to establish that the patterns of variable binding in language can *also* be viewed as cognitively structured patterns that comfortably align with the linguistic manifestation of variable binding. Here, it is important to distinguish cognitive constitution from a mere cognitive penetration/influence. The distinction can be best described by showing what is actually at stake when one merely delineates the cognitive scaffolding of various linguistic constructs or structures. For instance, van Hoek (1996) has argued that variable binding has an underlying cognitive reflex in the sense that a variable is bound by a (quantificational) binder if the binder is salient or prominent in the overall conceptual schema of the sentence(s). That is, the salient item is perceptually and

---

[1]Thus, for example, if we have the sentence 'Raj loves what he needs' where on one interpretation 'he' is bound by 'Raj', the logical analysis will be this: Raj [λx [x loves what x needs]]. It may be observed that the same lambda operator scopes over or controls the variable x which stands for the binder on the one hand and 'he' on the other.

thus cognitively foregrounded against the background projected by the variable or by the structure containing the variable. Thus, for example, in (61) 'the university' is the binder which is the salient item but 'its guidelines on the admission procedure' is the background item containing the variable that the binder binds.

However, the case in (62) seems to be problematic if these assumptions are carried forward, in that even though 'its guidelines on the admission procedure' is the salient constituent in the sentence, it contains the variable but not the binder. One way of getting around the problem is to argue that since 'its guidelines on the admission procedure' is the salient constituent in (62), this explains why variable binding cannot be established in (62) between the binder 'the university' and the possessive pronoun 'its.' That is, one may state that being salient is associated with the property of being a binder, but everything that is salient need not be a binder. What are salient in linguistic structures, after all, form a broader class than what binders are. But then, this weakens the force of the postulate that binders are salient structures in language because within the same linguistic constructions variables can also be salient. In other words, there is no unique way in which binders in linguistic structures are salient, whereas the variables binders bind or the structures in which they appear are expected to be less salient.

Whatever the merits or demerits of this proposal are, it is vital to recognize that this proposal rests on the principle that aspects of linguistic structures reflect aspects of our cognitive organization. Now, it is crucial to note that this does not guarantee that aspects of linguistic structures by virtue of reflecting aspects of our cognitive organization are *necessarily* constitutive of those aspects of cognitive organization. The issue concerned is a bit subtle, and hence, it requires a bit of spelling out. It needs to be emphasized that van Hoek's proposal is within the grammatical framework of Cognitive Grammar (Langacker 1987, 1999). Even though Cognitive Grammar in itself postulates that linguistic structures are intrinsically conceptualizations—conceptual schemas that are ultimately grounded in our sensory-motor systems—van Hoek's proposal in particular does not show that variable binding as a linguistic phenomenon is cognitively structured. Rather, it merely reveals the manifestation of aspects of our cognitive organization in

4 Linguistic Structures as Cognitive Structures

relevant linguistic structures by way of uncovering the relations between linguistic expressions that can be cognitively construed and hence reflect the imprint of cognitive scaffolding. Thus, this amounts to demonstrating that there are certain cognitive properties that underlie and thus underpin the structural relations in linguistic structures. But, unless one shows that the linguistically manifested properties and relations of a certain linguistic phenomenon are in a substantive sense properties and relations of the underlying cognitively structuring, what one ends up establishing is a weaker or diluted form of the cognitive makeup of natural language. Thus, for instance, in the case of variable binding if one just specifies the cognitively relevant properties of the binder and the variable it binds, it does not equate to showing the cognitive constitution of variable binding. In order to do what is desirable, one has to show how the complex ensemble of relations between the binder and the variable in natural language can be translated into an ensemble of relations that not merely turn out to be cognitively organized but also exhibit constraints similar to or the same as those found in natural language. This is what is implied when we deem that cognitive constitution needs to be distinguished from a mere reflection of the cognitive impact on aspects of language.

Equipped with this distinction, we may now try to find out what is really involved in variable binding that makes it a fundamentally cognitive phenomenon. To proceed further in this direction, we can ask the naïve questions: Why do binders bind (pronominal) variables? What does binding achieve in natural language? To be able to answer these questions, we can look closely at cases of variable binding to understand it better. When a binder (whether quantificational or non-quantificational) binds a pronominal, it transfers all or most of its content to the pronominal variable that lacks content. A pronominal is a variable exactly because it is gappy and requires content in order to contribute to the interpretation of the whole linguistic structure. Thus, a binder acts a kind of filler that fills up entities like variables. Unless the variables are filled up with contents, the whole interpretation of linguistic structures will remain gappy as well. This will definitely undermine the cognitive status of linguistic structures, for linguistic structures that continue to remain gappy in their interpretation are cognitively useless as they cannot be comprehended and further integrated into the overall texture of beliefs, ideas, concepts, and knowledge in language users. Binders by virtue of transferring their contents to variables endow variables with the status of full-fledged linguistic expressions that bear both linguistic form and linguistic meaning. What needs to be noted is that variables are not fully devoid of contents. Rather, they lack certain contents or linguistic meanings that can raise them to a level at which they can contribute to the interpretation of structural meanings. Many pronouns across languages carry features of gender, person, and number (e.g., 'he' vs. 'she' or 'we' vs. 'they' in English). Such contents may be called *formal* contents since it is precisely these contents that contribute to grammatically relevant operations such as agreement or the manifestation of argument structure in linguistic structures. So it appears that the content transfer operation from the binder to the variable must be such that it satisfies both the formal contents that are present as well as the semantic contents that are lacking in the variable. This explains why 'the father' cannot bind 'her' in (63) due to a mismatch in gender and 'the women' cannot bind 'her' due to number mismatch.

(63) The father of the children insulted her friend on the street.
(64) The women in blue dresses quite often met her sister during the summer.

It is evident that variable binding in the *intra-structural* context (within sentences or larger structures) is just one way of filling up variables that are by their very nature gappy. But this is not, of course, the only way variables can inherit contents from linguistic expressions other than themselves. Variables can inherit their contents from other entities that are either present in the discourse space of the speakers engaged in an interaction or imagined or somehow evoked *ex nihilo*. This is what we have observed in (58) where 'she' cannot be bound by 'Jane' and hence it has to be linked to some other entity which may have been contextually present in the discourse or has to be represented by way of imagination or evoked by analogy or something else. Clearly, this is not of the same sort as that which is found in the intra-structural context of variable binding. As a matter of fact, in the extra-structural context (as in (58)) the linking between a variable and the (potential) content filler is not a case of variable binding at all. While variable binding warrants a match of an ensemble of formal, structural, and semantic contents/features between the variable and the binder, a mere linking of the variable with a content filler outside structural context of the sentence(s) concerned links the semantic-pragmatic contents of the filler to the variable, in order that the variable comes to possess some representational content that can help track the indexical behavior of the variable. This is something quite different. Above all, this is an alternative to the standard operation of variable binding, and if they had been the same thing, they would not have identified alternative strategies of filling up variables with some kind of contents. Two alternative strategies of assigning contents to variables have disjoint linguistic contexts and spell out distinct cognitive consequences as well.

Notice that when a variable inherits the contents of a binder from within a linguistic structure, the contents of the binder merge or unify with those of the variable, whereas a variable that derives its contents from outside the structural context of sentences merely links to, or is paired with, another linguistic expression whose content features bear a measure of similarity along a number of dimensions to those of the variable in question. Let's call the former *content integration* and the latter *representational linking*. While content integration is more structurally constrained, representational linking is less structurally constrained and involves heavy cognitive resources and mechanisms. This difference will become clearer once they are explicated more precisely in cognitively significant terms. Content integration consists in unifying the content features of the binder with those of the variable, but representational linking pairs a whole representation of a linguistic expression of the content filler with the variable at hand. This can be formulated the following way.

(65) $A \& B \rightarrow A \cup B$ ----- Content Integration
(66) $X \& B \rightarrow (B, X)$ ----- Representational Linking

(Here, $A =$ set of content features of the binder, $X =$ set of content features of the content filler, $B =$ set of content features of the variable).

What (65) states is that two different sets containing content features of the variable and the binder are unified when variable binding obtains, whereas (66) says that the content features of the variable and the content filler become connected by virtue of the pairing of the variable with the content filler, that is, by virtue of forming a 2-tuple. It is quite clear from the formulations above that the order of the variable and the content filler matters in representational linking, but this is inconsequential in content integration. This is so because of some significant cognitive principles.

First, content integration merges the contents of both the variable and the binder in such a manner that a kind of *categorical invariance* is preserved in the unified entity constructed. Categorical invariance refers to the preservation of the content features either of the variable or of the binder in the merged construct that is formed when the content features of two different categories are fused into one. To put it in other words, no matter what the categorical features of the variable or of the binder are, they remain preserved in the fused construct that assumes the status of a cognitively configured unity. This unity is in a sense akin to the unity of distinct visual categories (texture, feel, luminosity, depth, etc.) into a single percept. This is not so in the case of representational linking. Representational linking can actually be thought of as the pairing of two distinct representations that the variable and the extra-structural content filler stand for. Since the extra-structural content filler is not a linguistic representation per se, it is re-presented as a linguistic representation for the purpose of the linking. Take the case of (58) again. The content filler that can fill up 'she' is not a linguistic representation per se; rather, some entity that can appropriately fit the contents of 'she' has to be represented in the mind (no matter how it is evoked), and then, it has to be (re)encoded in a linguistic format for the linking.[2] This particular content filler can (also) plausibly be in a multimodal representational format that is transduced into a linguistic format. The linking concerned ideally starts with the variable concerned,

---

[2]Even if the content-filler is present in some prior linguistic construction, that is, linguistically present in the discourse, it has to be (re)encoded in a linguistic format for the new linguistic context in which the variable in question is present.

and hence, the order of the items linked matters here. That is why representational linking is an ordered 2-tuple.

Second, while content integration is fine-grained and encompasses the component-like content features of the binder and the variable, representational linking is holistic and sometimes gestalt-like. Let's look at (55). The binding of 'he' by 'the guy over there' targets not merely the common content features of person, number, and gender of both the variable and the binder but also the syntactic-semantic features of both the items. The relevant syntactic-semantic features can be listed out below.

(67) 'the guy over there' knows P (where P is what one wants to do here)
'he' wants to do something here (where what he wants to do here is P).

When the pronominal is bound by the binder, these syntactic-semantic features along with the features of person, number, and gender bond together and merge in such a way that the direction of emergence of the features concerned does not really matter. What matters is whether or not all the relevant features rather than only a subset of them have been fused into an integrated representational whole within which the content features either of the pronominal variable or of the binder are not to be isolated or distinguished. On the other hand, representational linking *by default* starts with the variable at hand and then places it in order with the pertinent content filler that is evoked from some appropriate context and so represented. That this is bound to be of this form is because representational linking involves representational selection which consists in selecting the most appropriate representation that has an optimal fit with all the constraints the syntactic-semantic features of the pronominal impose on the context. Thus, one representation which is already available serves to act as a kind of 'model' of the other representation to be searched for and selected. Once the selected representation is matched with the already available representation, the pairing is consolidated and the two representations are appropriately connected to one another.

The distinct cognitive constitutions of variable binding and coreference between a variable and its content-filler are responsible for the linguistic patterns that are observed in language. Note that representational linking is cognitively marked (by virtue of being heavy on cognitive resources and mechanisms) but structurally unmarked (in involving less linguistic materials or structural details because the linguistic structure of the content-filler remains absent), whereas content integration is structurally marked (as the structural details of both the binder and the variable are to be taken into account and then put together) but cognitively unmarked (because the linguistic representations of both the binder and the variable are available, and they can be immediately mapped onto their associated mental representations). Given these differences between them, it would be interesting to see what would happen if there are multiple binders or variables in a single linguistic structure. Let's then look at the examples below.

(68) Although Bob knows the person who has influenced Rocky, he does not want to discuss this with him.
(69) The driver came with a boy and he asked him to wait outside.
(70) As she came closer to him near the table where they used to chat, he hurled himself upon Tina and saw Rina right there.

In (68), the salient binder for 'he' is 'Bob' and this is due to the fact that 'Bob' is the 'topic' of the sentence which anchors the sentence in its informational-structural context, thereby embracing content integration since the topic status of 'Bob' naturally allows for content integration by making 'Bob' structurally attractive. But the pronominal 'him' can be bound either by 'the person...' or by 'Rocky' or even by an extra-structural content-filler. Now if the choice is between 'the person...' and 'Rocky,' the binding relation between 'him' and the exact binder has to be determined by syntactic-semantic relations between the possible binders and the variable. This determination is not like the resource-heavy cognitive search as is valid in the case of representational linking. Besides, content integration does not *in itself* involve the selection of a binder from within the structural domain of the linguistic expression concerned. Rather, it refers to the post-selection merging or unification of feature contents. Given the distinction between content integration and representational linking, what we can predict is that when there are binders available within the structural domain of a given linguistic expression, an option that meshes well with the (linguistic) structural organization of representations is to be preferred as the default option because it can tie the structural representation of the binder to that of the variable within a restricted domain of the linguistic structure. Hence, variable binding in (68) is to be preferred to a co-reference between 'him' and a certain extra-structural content-filler.

There are also other specifically linguistic constraints that govern how variable binding is expressed in language. There are well-known linguistic differences between quantificational variable binding and non-quantificational variable binding. Quantificational variable binding involves a quantificational binder, whereas non-quantificational variable binding involves any other noun phrases as binders. While quantificational variable binding is not generally sensitive to clausal boundaries, non-quantificational variable binding can be. The following examples illustrate the distinction (see Baltin et al. 2015).

(71) a. Jón   sagði að   eg   hefði   svikið   sig
         John said that 1PERS-SING had   betrayed 3PERS-SING
         'John said that I had betrayed him.'
     b. Jón veit   að   pu   hefði   svikið   sig
         John knows that 2PERS-SG have   betrayed 3PERS-SING
         'John knows that you have betrayed him.'   (Icelandic)
(72) a. Every woman believes that the man who loves her is a prince.
     b. Every woman told every man that people knew that she
         should leave.
                        (PERS = person, SING = singular number)

The Icelandic examples in (71) show that non-quantificational variable binding is *local* in the sense that 'Jón' cannot bind 'sig' across clausal boundaries in (71b) but it can in (71a). The nature of the matrix predicate is also important here as the matrix predicate in (71a) is a non-factive predicate, that is, it presupposes merely a statement which need not be a fact, but the matrix predicate in (71b) is a factive predicate which presupposes a fact in its embedded clause. This does not, of course, constrain quantificational variable binding as shown in the English examples in (72).

Now if content integration is way distinct in its cognitive constitution from representational linking, this particular division in the behavior of variable binding must be attributed to the very nature of content integration rather than to representational linking. In fact, this variation in the behavior of variable binding follows straightforwardly from content integration. Recall that content integration is constrained by the nature of linguistic representations in particular, and that it is linguistically marked in involving comparisons and analyses of greater structural details. Since content integration by virtue of the operation of set union (see (65)) is not in itself domain-dependent, it is underspecified with respect to the *demarcated* linguistic domain (clausal or trans-clausal) within which content integration may apply. Hence, content integration may be specified in different ways in different sizes of linguistic domains. But note that this does not hold true for representational linking, since the relevant content-filler does not even exist in a domain that is either linguistically constructed or linguistically restricted, and hence, its representation and the variable concerned remain in disjoint domains that cannot be immediately coalesced into a single connected domain for domain measurements. Even when a content-filler is present in some prior linguistic context within a given discourse, its linguistic representation has to be culled from a set of prior sentences forming the linguistic context in which the content-filler exists, which makes it all the more difficult to demarcate the linguistic domains which may vary in terms of certain parametric specifications. The relevant content-filler is always outside the sentence in which the variable is located. This makes it impossible to define parameters that specify intra-sentential and extra-sentential categories, precisely because the content-filler has to be *at all times* extra-sentential. Besides, clausal boundaries within a complex sentence are easy to demarcate, and hence, it is also easy to fix the relevant specifications, but it is not in the same way easy to fix the specifications across sentential boundaries. In the case of clausal boundaries within a complex sentence, the linguistic significance of an expression may depend on whether it is within the matrix clause or outside of it, whereas if some linguistic expression is always outside a sentence, its significance cannot be so determined because its presence within the boundaries of the relevant sentence has never been gauged. This explains why the absence of linguistic material in fragmented linguistic structures does not block the retrieval of the linguistic representations for variable binding.

(73) The tailors make their children work, and so do the cobblers.
(74) The tailors make their children work. The cobblers do, too.
(75) They make their children work. The tailors make their brothers work. The cobblers do, too.

It is clear that in (73) 'the cobblers' binds 'their' as it is retrieved from the linguistic representation in the previous conjunct 'the tailors make their children work.' What is interesting in (74) is that variable binding is still possible even when the fragment expression containing the binder 'the cobblers' forms a separate yet dependent sentence. Note that the sentence 'The cobblers do, too.' is not an independent sentence even though it appears independently right after the matrix sentence from which it derives its own linguistic representation for variable binding. But (75) shows that the dependence of the fragment sentence on the preceding matrix sentence breaks down when it is interrupted by an intermediate sentence interpolated in-between. That is, in (75) 'the cobblers' does not bind 'their' from the linguistic representation of the sentence 'They make their children work' but from 'The tailors make their brothers work.' Thus, (75) means that the cobblers make their brothers (but not the children) work. Additionally, on a natural reading of (75) 'they' in the first sentence of (75) cannot be bound by 'the tailors' or by 'the cobblers.' For 'they' to get its content from some linguistic referent, there has to be some preceding sentence containing the linguistic representation of the binder—which makes the entire linking process extra-clausal or extra-sentential and thus not straightforwardly domain-bound. But the absence of some preceding sentence with the linguistic representation of the binder, as it is in (75), renders representational linking domain-independent. This reveals that there are constraints on the extent to which extra-sentential elements can have an impact on elements within other preceding or following sentences. The retrieval of the appropriate linguistic representations for variable binding is possible in the immediate vicinity of the following sentence, but not in the reverse direction (i.e., in the preceding sentences). Hence, content integration is tightly coupled to the linguistic constraints that modulate or govern the ways in which linguistic representations are made available for variable binding, whereas representational linking does not work that way as it cannot automatically go up or down the connected path of sentences for the exact content-filler. Overall, this reinforces the underlying idea that linguistic structures are themselves cognitively constituted and constrained by the nature of the cognitive substance. We shall now examine certain systematic linguistic connections between some quantificational elements that are indicative of their cognitive constitution.

4.1.2 Quantifiers

Natural language makes a linguistic distinction between exhaustive/inclusive choices, freedom afforded in making choices and a selection from among a set of choices. These concepts are reflected by 'all,' 'any,' and 'only,' respectively, in English. Although different languages may have different ways—morphological, syntactic, phonological, or a combination of these—it is evident that natural languages across regions/linguistic families must have some means of expressing these concepts. A language lacking any formal device(s) for these concepts may be less expressive than the one that incorporates them. The conceptual distinctions between 'all,' 'any,' and 'only' equip natural language users with very powerful expressive devices that reflect and also realize finer distinctions in meaning relations constructed within the cognitive realm. The following sentences make this clearer.

(76) All students have left.
(77) I want all the articles on structuralism.
(78) Any car will do for now as far as she is concerned.
(79) Only our politicians get all the advantages—others are disregarded.
(80) Any child can break the bar, but only you cannot.

What is palpably clear here is that the sentences (76–77) make certain meaning distinctions that are not to be found in (78) or (79). While (76–77) talk about the entire class of a group of things within the universe of objects that bear a certain relation designated by the structures minus the quantificational phrases, (78) or (79) isolates only a portion of some demarcated region of the universe of objects that possess some characteristics conveyed in the pieces of structure minus the quantificational phrases. The sentence in (80) is a bit different because it combines the properties of both 'any' and 'only.' There are in fact subtle meaning differences between 'any' and 'only' that can be easily contrasted with the meaning of 'all.' Note that 'all' denotes the inclusion of a set of entities, and if this is the case, the relation that obtains between the quantificational noun phrase containing 'all' and the predicate is that of inclusion too (see Barwise and Cooper 1981; Peters and Westerståhl 2006). Thus, in (76) the set of students is a subset of the set of all those individuals who have left. The story of 'only' is a bit different. 'Only' when used before a noun identifies a specific set from among a domain of other sets, and hence, the set designated by the predicate is a subset of the set denoted by 'only.' In this sense, 'only' designates a relation which is the reverse of what 'all' designates. Thus, for example, in (79) the set of all individuals that get advantages is a (*not necessarily* proper) subset of the set of our politicians. The relation that 'any' gives rise is way distinct from that of either 'all' or 'only.' 'Any' is a bivalent quantifier in its behaviors, in that it sometimes behaves like 'every' and sometimes as 'some.' The distinction concerned is well expressed by the difference between (78) and (80). It is easy to notice that (78) talks about a non-specific car in the sense that the speaker's expression does not make any commitment with respect to the choice of a particular car; rather, the speaker makes it clear that some car of any make or type or quality as appropriate in the context will do as far as the compliance with another female person's taste or preference is concerned. On the other hand, on the preferred reading (80) does not mean that there is a certain child who can break the bar but rather that every child can break the bar. That is, in (80) 'any' has a *distributive* interpretation of 'all' which has a *collective* reading as well. Consider the following cases.

(81) All my books are lost.
(82) All the guests listened to the performance of the orchestra in utmost amazement.

While (81) means that every one of my books is lost (the distributive interpretation), (82) implies that all the guests in unison listened to the orchestra performance in utmost amazement (the collective interpretation). This shows that 'all' has a duality in its interpretative behavior. From this perspective, there is reason to suppose that the behavior of 'any' parallels that of 'all.' 'Any' has a distributive reading just like 'all,' and the reading of 'any' on which it implies a (single) non-specific entity can be aligned with the collective reading of 'all' *only if* we assume that the collective entity is analogous to the singular non-specific entity. But this may come as a surprise, given that 'any' after all designates freedom in choice while 'all' designates inclusive choice. What is more interesting is that the behavior of 'any' also parallels that of 'or'—an entity which is not, of course, a quantifier. Just as 'any' can have a distributive reading of 'every' along with an *existential* reading (the existence of some entity specific or non-specific), the disjunctive linker 'or' can have both exclusive and inclusive readings. The following examples illustrate this well.

(83) He has not understood either his wife or his daughter.
(84) Perry or Das can tackle this problem.
(85) You can take tea or coffee.
(86) Take a train or a bus to go to school.
(87) There are plenty of books or papers on this delicate matter that you can read.

As one examines (83–86), it becomes amply clear that 'or' does not have a uniform reading all throughout. Whereas (83–84) have a distinctive inclusive reading on which the two individuals separated by 'or' are included in the meaning relation structurally conveyed, (85–86) segregate one entity from the other in terms of a selection from a pair of choices, thereby instantiating the exclusive reading. The curious fact that 'or' can have both inclusive and exclusive readings makes

4 Linguistic Structures as Cognitive Structures   161

for a subtly complex scenario in (87) in which 'or' appears to lead to a mixed interpretative possibility. On the one hand, (87) definitely has an inclusive reading whereby the speaker intends to convey the impression that both books and papers on the delicate matter at hand are available. But, on the other hand, it seems that the speaker is sure that there are plenty of materials on the delicate matter available but not sure whether they are books or papers. Thus, the speaker leaves the matter of selection open for the recipient of the utterance to decide about. From this perspective, the distributive reading of 'any' seems concordant with the exclusive reading, in that the exclusive reading isolates *each* of the choices from the other just like the distributive reading of 'any.' Analogously, the existential reading of 'any' is orthogonal to the inclusive reading on the grounds that the inclusive reading underlines the *inclusive existence* of a set of entities that does not involve making a choice from a selection. As the existential reading consists in specifying the existence of some entity or entities, the inclusive reading characterizes the existence of two or more entities to be taken into consideration together. For instance, the example in (84) does not simply assert that both Perry and Das possess the skill or capacity to tackle the problem; rather, it also emphasizes that there exist two persons, namely Perry and Das who can tackle the problem in question.

At this point, it is worth checking if 'only' distinguishes itself from both 'any' and 'or' since it is the inverse of 'all.' That is, because 'any' is patterned on 'all' and 'or' seems to be orthogonal to 'any' on several counts, the behavior of 'only' can be expected to be inversely distinct from that of both 'any' and 'or.' This warrants scrutiny. To this end, we may go through the following examples.

(88) Only children are allowed inside this park.
(89) Only Jeet can understand the prevalent condition in the village.
(90) In the restaurant, he ate only chocolates.

In (88–90), the behavior of 'only' is manifested in its remarkable *monovalent* character that does not evince the kind of duality 'any' and 'or' exhibit. That is to say that 'only' uniformly scoops out the focused individual or collection of entities from the meaning relation expressed by the rest of the sentence. Thus, for example, in (88) 'only' elevates the set of children to a space within which all those who are allowed inside the park are included. This is another way of saying that all those other than children are bracketed off by 'only' in (88). Similarly, in (89) all individuals other than Jeet are kept out of the focus domain demarcated by those who can understand the prevalent condition in the village, and in (90) everything else other than chocolates is what is backgrounded. In all cases, the relation between the set characterized by 'only' and the one identified by the rest of the sentence is reversed with respect to the relation 'any' implements. Hence, in (89) the set of all those who can understand the prevalent condition in the village is included in or is equivalent to the set containing 'Jeet' alone, and in (90), the set of all items eaten by the male person in the restaurant is included in or identical to the set of chocolates. On the universal reading of 'any,' the set specified by 'any' is included in the set specified by the structure minus the phrase containing 'any'; thus, in the sentence 'Any child can do it.' the set of children is included in the set of all those who can do it.

Now, what is crucial for us to note that the relation of inclusion 'only' implements is unique in many ways. 'Only' does not always incorporate in its semantic representation exhaustive inclusion. Rather, it may favor partial inclusion as well—which makes the character of the inclusion kind of *leaky*. This is remarkably evident in the following cases.

(91) Only parents of the kids roam freely in the backyard of the school.
(92) Only professors of cognitive science use complicated machines to study the mind.
(93) This place has seen only Italian visitors.

Notice that (91) does not, strictly speaking, say that those who roam freely in the backyard of the school are exactly the individuals who are parents of the kids. That is to say that it is not really necessary that the set of individuals who roam freely in the backyard of the school is identical to the set of individuals that are parents of the kids. This leaves room for the possibility that there could be some parents of the kids who do not roam freely in the backyard of the school. What matters for the satisfaction of the semantic conditions of 'only' in (91) is whether or not those who roam freely in the backyard of the school are *all* parents of the kids. Thus, if it turns out that someone other than the parents of the kids *also* roams freely in the backyard of the school, then it decidedly violates the semantic condition implicit in 'only,' on the grounds that at least someone has been found who roams freely in the backyard of the school but is not a parent of the kid(s). Likewise, in (92) if anybody other than professors of cognitive science uses complicated machines to study the mind, this will incontrovertibly do away with the inherent meaning condition that 'only' warrants. And similarly, in (93) if visitors other than those who are Italians are to be found to have been to the place concerned, this will again nullify the inclusion relation that 'only' requires in a unique manner.

Hence, it is quite clear that the inclusion relation of 'only' can allow for partial inclusion, even though the canonical or prototypical inclusion relation 'only' licenses is that of exhaustive inclusion. This contrasts with the inclusion relation 'all' warrants in another significant way, in that the inclusion relation that 'all' licenses is canonically that of partial inclusion. Hence in example (76) above the set of students is *partially* included in the set of those who have left. This is so because it is perfectly possible that not everyone who has left is a student, and there could be someone who has left but does not belong to the group of students. As it is not necessary that 'all' must always instantiate partial inclusion, 'all,' just like 'only,' allows for a less canonical condition within which it permits exhaustive inclusion. Consider, for example, 'All triangles have three sides' where the set of all triangles is equivalent to the set of all things that have three sides and vice versa. There is no reason whatsoever to doubt that 'all' and 'only' are inverses of each other, but what is more significant here is that 'only' does not abide by a general constraint on natural language quantifier interpretations which is uniformly obeyed by all the standard quantifiers including 'all.' This is called the Conservativity Constraint (see Keenan and Stavi 1986), which conserves the *domain* set (which is characterized with respect to the noun head attaching to the quantifier; for example, the set of triangles in 'All triangles have three sides' is the domain set) over the *restriction* set (the set characterized with reference to the predicate (head) of the construction; thus, for example, the set of all entities having three sides in 'All triangles have three sides' is the restriction set). Mathematically, for the quantifier 'all' this can be expressed as $Q(D)(R) = D \subseteq R = D \subseteq (D \cap R)$, where $D =$ the domain set, $R =$ the restriction set and $Q =$ 'all.' For the example of 'all' in (76), this can be expressed as 'all students are students that have left'—note that this *re-expresses* the set of students over the set of those who have left and hence $D \cap R =$ the set of students that have left. The domain set is thus iterated over the restriction set and so conserved even within the restrictions specified by the restriction set.

One needs to be careful to understand the import of the Conservativity Constraint in the linguistic behavior of natural language quantifiers. Even though this constraint appears to bind together all human language quantifiers, 'only' when used in its quantificational form[3] and 'many' seem to invalidate the constraint in not conserving the equivalence between the non-conservative relation and the conservative relation. For instance, the conservative version of (91)—that is, 'only parents of the kids are parents of the kids that roam freely in the backyard of the school'—does not have the same logical structure as the non-conservative version found in (91). While the non-conservative version is a contingent statement that says that none other than the parents of the kids roams freely in the backyard of the school, the conservative version turns out to be a tautology because the set of parents of the kids that roam freely in the backyard of the school cannot be smaller in size than the set of parents of the kids (this is what obtains in partial inclusion). If we suppose that the set of parents of the kids that roam freely in the backyard of the school is smaller in size than the set of parents of the kids, then there must be some parent(s) of the kid(s) that do(es) not roam freely in the backyard of the school. But this is

---

[3]Many people do not consider 'only' to be a standard quantifier because it can be used before a whole (quantificational) noun phrase itself as in 'only the president', 'only three soldiers', 'only Meera' or even in 'for students only'. If 'only' can precede a quantifier itself in a (quantificational) noun phrase, this indicates that 'only' is not located in the position of quantifiers in (quantificational) noun phrases. This is not possible for standard quantifiers such as 'every', 'a/some', 'most' etc. (see also von Fintel 1997). Besides, 'only' can also be used adverbially as in 'He only pushed the bar', and as an adjective in examples like 'the only man', 'my only book'.

logically impossible since the augmented restriction (D ∩ S = the set of parents of the kids that roam freely in the backyard of the school) has already specified that the parents of the kids roam freely in the backyard of the school, and if so, it is nothing more than a contradiction to state that there are some parent(s) of the kid(s) that do(es) not roam freely in the backyard of the school. Therefore, what we have here is this: D = D ∩ R, and so Q (D) (R) = D ⊇ R = D ⊇ (D ∩ R) = D ⊇ D = (D = D), where Q = 'only.'

The quantifier 'many' tells a quite different story. On the one hand, 'many' does not have a *poly-categorial* character that 'only' has (in manifesting properties of grammatical categories of determiners, adverbs, and adjectives), it is a proportional quantifier that discernibly overturns the Conservativity Constraint. If, for example, we have the sentence 'Many books in the library are on thermodynamics,' the proportional interpretation that 'many' gives rise to will dictate that the proportion of books on thermodynamics in the library must be greater than that of other books in the library. Mathematically, this can be expressed the following way.

(94) $Q(D)(R) = \dfrac{|D \cap R|}{|D|} > \dfrac{|D \cap R^C|}{|R^C|}$

The formulation in (94) states that the proportion of the set of books in the library that are on thermodynamics (when divided by the size of the set of books in the library) must be greater than the proportion of the set of books in the library that are not on thermodynamics (when divided by the size of the set of entities that are not on thermodynamics).

What makes the behavior of 'many' orthogonal to that of 'only' is how it nullifies the Conservativity Constraint just like 'only' but in a somewhat different way. That is, unlike 'only,' the quantifier 'many' does not exactly lead to a tautology when the conservative version is compared to the non-conservative version; rather, the conservative and non-conservative versions yield to varying context-sensitive interpretations. Consider the examples given below.

(95) Many soldiers die on the war front.
(96) Many countries waver between violence and peace.

Now it is evident that the conservative version of (95)—that is, 'many soldiers are soldiers that die on the war front'—does not retain the same proportional interpretation in the contingent satisfaction condition of circumstances that the non-conservative version possesses in (95). That is to say that the conservative version would be true or simply valid in circumstances in which the non-conservative version would not. Imagine a scenario in which a battalion that usually keeps to the number of 1000 soldiers often goes to the war front, and around 100 soldiers are generally found to die on the war front. In such a situation, the conservative version would be legitimate as the interpretation of 'many' targets some fixed estimate of the number of soldiers that counts as 'many soldiers.' Thus, we say many soldiers (around 100 as per the fixed estimate) are soldiers dying on the war front. But in such a scenario the non-conservative version *may* not be true since 100 out of 1000 soldiers dying on the war front may not be a good measure of a higher proportion licensed by the proportional reading of 'many,' simply because the proportion of those soldiers that die cannot certainly be higher than that of those soldiers that do not.[4] This is because the conservative version turns out to produce a *cardinal* (an estimate of some fixed number) interpretation of 'many,' whereas the non-conservative version has a general proportional reading, as specified in (94).

The same point pretty much holds true for (96) as the conservative version of (96)—'many countries are countries that waver between violence and peace'—is rendered cardinal. The reason is again very simple. In (95), for instance, we do calculate a proportion that places the set of countries that waver between violence and peace in relation to those countries that do not do so. But when the conservative version is

---

[4]This, of course, depends on the measure of $|R^C|$ which is, in the present case, the *cardinality* of the set of those individuals who do not die on the war front (see (94)). The value of this may vary depending on the context. For instance, if the value of $|R^C|$ is 10,000, then the calculation of the values taken from this example in terms of the equation of (94) will produce the desired inequality. However, if, for example, the value of $|R^C|$ is 6000, the desired inequality will not obtain. In other words, in this case 100 soldiers dying on the war front will not count as 'many soldiers'.

derived from it, the interpretation of 'many' implicitly imposes a sort of bound (intuitively pre-decided) on the number of countries that waver between violence and peace, whereas the non-conservative version demands that the proportion of countries wavering between violence and peace as individuated by 'many countries' be higher than the proportion of those countries that do not do so. This inequality of proportions, to put it succinctly, does not accord with the fixed bound interpretation in the conservative version. This decidedly spoils the possibility of maintaining a logical equivalence between the conservative version and the non-conservative version, as warranted by the Conservativity Constraint. All quantifiers that obey the Conservativity Constraint preserve the logical equivalence between the conservative version and the non-conservative version, as discussed above for quantifiers like 'all.'

Given this recalcitrance in the behavior of 'only' and 'many,' it behooves us to ferret out the reasons for the bizarre behavior of these two words. On the one hand, one may not, of course, make much of the quantificational character of 'only,' insofar as it exhibits many non-quantificational features that are, in some respects, discordant with the character of quantificational elements. And on the other hand, one may attempt to brush off the case of 'many' by having it relegated to the category of trivialities which are determined by context-sensitive factors and properties. In the face of these considerations, it appears that one need not really bother about the anomalous character of both 'only' and 'many.' However, it must be emphasized that this line of reasoning is itself deviant and misguided on the grounds that the aberrant behavior of 'only' and 'many' needs to be explained rather than explained away or even stated as simply exceptions. If the explanatory aim of the present work is to show how linguistic structures can themselves be marked as cognitive structures, it is necessary to figure out why 'only' and 'many' override the Conservativity Constraint.

If we look closely at the behavior of different kinds of quantifiers in natural language, some logically interesting patterns conspicuously strain out of the space of otherwise well-grounded differences among them. The Conservativity Constraint is one such general logical constraint on all types of natural language quantifiers that governs the space of all possible quantifiers linguistically realizable (see also van Benthem 1983). But then we face the problem of accounting for the exceptions that 'only' and 'many' lead to. One solution that has been proposed to bring all natural language quantifiers under the ambit of a general constraint takes recourse to another generalization on the logical patterns of quantifiers. This is called the Witness Set Constraint which consists in characterizing a witness set over the *domain* set in such a way that a subset (proper or otherwise) of the domain set is preserved over the relation between the domain set and the restriction set. The idea of witness sets goes originally to Barwise and Cooper (1981), but its usefulness in defining a constraint on the class of natural language quantifiers has been pointed out in Fortuny (2015).

The underlying idea here is that the exact type of relation between the domain set and the restriction set will remain invariant even when we define the same relation between *any subset* of the domain set and the restriction set. Thus, for example, in the sentence 'Some cakes are left in the kitchen' the witness set for the domain set (which is a set of cakes) will be a non-empty set of cakes, and this set turns out to be a subset of the entire set of cakes and also to have a subset relation with the set of things that are left in the kitchen (as well as having an intersection with it). This is sufficient for a quantifier like 'some' since it permits inferences from subsets to supersets, for if some cakes are left in the kitchen, it must also be true that some cakes are left. That is, the set of things left in the kitchen is a subset of the set of all things left (somewhere), thereby regimenting an inference from a subset to a superset. This pertains to a property called *monotonicity*. Hence 'some' is *monotonicity-increasing* quantifier.

When translated in these terms, 'only' seems to abide by this constraint. Let's see how. Let's take the same example that we discussed earlier—that is, the example in (91). If it is the case that only parents of the kids roam freely in the backyard of the school, the restriction set (i.e., the set of those who roam freely in the backyard of the school) must be a subset of the set of parents of the kids. We are aware that this is the typical property of 'only.' Now the witness set extracted from the domain set (i.e., the set of parents of the kids) will be any set of parents of the kids. Since 'only' is neither monotonicity-increasing nor monotonicity-decreasing (in permitting inferences from supersets to subsets, such as 'few'), an intersection between the restriction set and the domain set must be equal to the witness set (Fortuny 2015). That is to say that the relevant set of parents of the kids that roam freely in the backyard of the school must be equivalent to any (sub)set of parents of the kids. Likewise, for the quantifier 'many,' the proportional reading should be preserved when the witness set is defined. Thus, for instance, for the sentence in (95) the proportional relation between the set of soldiers and the set of those who die on the war front must be intact when the witness set is characterized as any set of soldiers that contains many of them. This is another way of stating that the proportional relation as formulated in (94) will remain intact when the relevant witness set is any set of soldiers containing many soldiers that can be reckoned to be a subset of the entire set of soldiers.

As we ponder over this, this turns out to be intuitively correct. But one may wonder why this should be the case. What cognitive consequences does this carry? Or even does this have any cognitive import at all? Obviously, witness sets are ways of conserving some part (or the whole[5]) of the domain set over the relation the domain set constructs with the restriction set. In other words, witness sets carry over a part of the domain set to the relation that the domain set enters into along with the restricted set. This indicates that the Witness Set Constraint is a more general principle of conservation of (parts of) domain sets. Although the Witness Set Constraint seems to render the Conservativity Constraint redundant by virtue of the fact that it constitutes a more general alternative that comports with the behavior of all natural language quantifiers, it is important to underline the need for an analysis that reveals the cognitive structuring of the relevant patterns of behavior that we find in natural language quantifiers. The nature of the cognitive structuring involved will thus tell us how natural language quantifiers including 'only' and 'many' are encompassed by the Witness Set Constraint but not by the Conservativity Constraint. Put succinctly,

---

[5]This is so because one of the fundamental axioms of set theory is that a set is a subset of itself. Hence any set *as a whole* is a subset of itself.

the specific cognitive structuring to be unearthed will help understand why the linguistic behaviors of 'only' and 'many' are fully compatible with Witness Set Constraint but are not adequately described by the Conservativity Constraint.

With this in mind, let's proceed to show how the linguistic behaviors of 'only' and 'many' naturally fall out of the cognitive structuring that underpins their linguistic structures. What is going to be suggested is that all quantifiers have a basic logical format structured around exclusion/inclusion relations. These relations can be uniformly described in terms of certain properties of set-theoretic structures having a deeper cognitive grounding. The relevant idea is that quantifiers manipulate exclusion/inclusion relations in ways that reflect shifting patterns in the cognitive mapping of conceptual relations. Sometimes some conceptual relations are structured in such a way that a given structural space is foregrounded with the concomitant backgrounding of another structural space that constitutes what is excluded, and sometimes two structural spaces—one foregrounded and the other backgrounded—occur in parallel and co-exist with one overlapping the other. The intricate dance between such structures characterizes the patterns that define the linguistic behaviors of quantifiers. First of all, it is significant to realize that the set-theoretic relations of a quantifier are *identical to* the inclusion-exclusion relations of the quantifier concerned. This can be illustrated with suitable examples from quantifiers. Thus, for instance, take the case of 'all,' which is known to include domain set within the restriction set in a manner that indicates that a part of the restriction set *may* be de-focused with the domain set constituting the foregrounded space in the mental organization of conceptual relations. Now what is interesting to note here is that since 'all' generally licenses partial inclusion, the canonical relation obtaining between the domain set and the restriction set is that of partial conceptual inclusion. The order of the inclusion is very significant here, in that the conceptual relation in the quantifier 'all' has a direction of inclusion from the domain set to the restriction set, which determines which portion of what is to be foregrounded or backgrounded.

With this in place, we may now wonder what exactly we mean by a structural space that has a cognitive grounding. Once the relevant notion of this space is clarified, the nature of the cognitive structuring will become clearer. A mentally constructed structural space corresponds to a set-theoretic space that encloses of a set of entities which could be either a delimited *partition* of a set or simply the whole set itself. That is, on this formulation a cognitively structured space is identical to a set of entities that delimits a portion within a set-theoretic space (represented in terms of Venn diagrams, for example). Thus, for example, if 'all' includes the domain set within the restriction set, this gets the cognitively structured space consisting of elements from the domain set enclosed and foregrounded within another cognitively structured space defined by the restriction set. The part of the restriction set that is excluded here constitutes a cognitively structured space that forms the 'contour' of the cognitively structured space characterizing the foregrounded space. The relevant notion of a cognitively structured space has a reflex of *mental spaces* as formulated in Fauconnier (1994). Since mental spaces are also set-theoretic objects that can entertain certain conceptual relations, cognitively structured spaces are, in this sense, akin to mental spaces. At this juncture, it is crucial to understand that the excluded cognitively structured space bears a certain relation with what is included. In fact, this relation can be formulated in terms of the notion of convex/non-convex regions. Convex regions are those in which for any two points (say, x and y) in a given region (say, R in some Euclidean space) all points that can be plotted on that line are also in R. The following diagram (with the relevant lines connected by points A and B) shows this quite clearly (Fig 4.3).

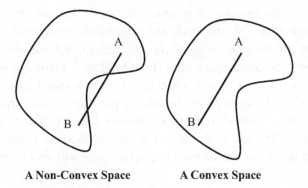

A Non-Convex Space   A Convex Space

Fig. 4.3 Non-convex and convex spaces

This idea has been incorporated as part of the formalization of cognitive spaces that have certain well-defined geometric properties organized in terms of a variety of dimensions defined by the properties of a given domain. The color domain, for instance, is defined by dimensions of hue, shade, reflectance, etc. These cognitive spaces have been called *conceptual spaces* which can have convex regions when the fact that some objects located at any two points with respect to a conceptual space instantiate a certain property licenses the conclusion that all objects located between those two points also instantiate the property at hand (see Gärdenfors 2000).

Equipped with this understanding, we may recognize that the excluded cognitively structured space in the case of 'all' in particular forms a convex region with respect to the cognitively structured space that is foregrounded. In other words, when the quantifier 'all' gets the domain set (partially) included within the restriction set, what happens is that the cognitively structured space corresponding to the domain set, which now corresponds to a partition of the restriction set, forms a convex region with reference to all points that can be marked on any line drawn all over the region comprising both the focused cognitively structured space and the excluded cognitively structured space which is a part of the restriction set. Since the cognitively structured space that corresponds to the partition of the restriction set excluded enters into a convex relation with respect to the cognitively structured space that is focused, any line drawn over the whole connected region comprising the two will preserve convexity, as shown in Fig. 4.4.

This is simply because all points that can be charted on a line connecting one point in the excluded cognitively structured space and another in the included cognitively structured space will decidedly fall within the connected space that includes the domain set within the restriction set. It may also be observed that the focused cognitive space constituted by the domain set forming a partition of the whole restriction set forms a convex region in a similar fashion. If the two points making up a line are both chosen from the focused cognitive space, the overall region of the focused cognitive space will also have a convex

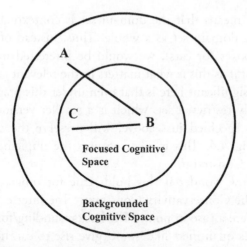

Fig. 4.4 The convex region in the connected space of 'all'

character. This is what is also shown in Fig. 4.4 where the points C and B connect a line other than the one linked by A and B. Importantly, the quantifier 'all' transfers the domain set *in toto* to the restriction set in all circumstances, and hence, the focused cognitive space must preserve the *intersection* of the domain set with the restriction set, which is identical to the domain set itself. This is another way of saying the focused cognitive space remains invariant over all possible re-configurations between the domain set and the restriction set in the case of 'all.' Readers may check that this holds true even in cases where the domain set of 'all' actually shrinks under certain pragmatic pressures, as in the following.

(97) All boxes are now empty.
(98) All cars have left the expressway.

Clearly, under ordinary interpretations of 'all' the sentences in (97–98) do not state that all boxes (or cars) in the universe are now empty (or have left the expressway). Rather, the relevant domain set defined by 'all' is here restricted to the set of boxes (or cars) *in the given context of utterance*. This means that the domain set is contextually restricted to a subset of the domain set as a whole. Thus, instead of talking of the whole set of boxes (or cars), we would be interested in talking about only a subset of it as this is what matters in the relevant context of utterance. What is significant here is that even under this pragmatic pressure the contextually restricted set, which is a smaller version of the whole domain set as specified just above, will preserve the same convexity as shown in Fig. 4.4. This is what ensures the implementation of the Conservativity Constraint.

One may now wonder if this holds true for other quantifiers that comply with the Conservativity Constraint. The intersective quantifiers like 'a,' 'some' are a case in point. The backgrounding/foregrounding of cognitive spaces quantifiers like 'some' give rise to can help understand the cognitive structuring of quantificational linguistic structures in a distinctive way. It is easy to recognize that quantifiers like 'some' establish a non-null intersection between the domain set and the restriction set (see Barwise and Cooper 1981). That is, if we say 'some children are naughty,' we mean to say that there must be *at least* two individuals who are children and are naughty. Another way of stating this is to say that the set of individuals who are children and are naughty cannot be empty. Then what forms the focused or foregrounded cognitive space is constituted by parts of both the domain set (the set of children, in the example) and the restriction set (the set of those who are naughty), and the rest is the excluded backgrounded cognitive space.

As is evident in Fig. 4.5, the focused cognitive space in the middle of the two circles constituting the intersection between the domain set and the restriction set forms a convex region with the backgrounded cognitive space. That is why the lines linked by A and B and by B and C both fall within the same connected region. In this respect, 'some' is just like 'all.' What, however, differs in the picture here is that the excluded cognitive space cannot in itself establish a convex region since it is not the case that for all points x and y within this region any two points between x and y will also be within the same region. This is shown by means of the line connected by D and E in Fig. 4.5—there are points on this line that fall outside the region of the connected circles. It is clear that the focused cognitive space in this case will also remain

4 Linguistic Structures as Cognitive Structures

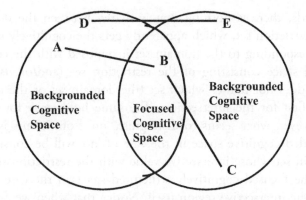

Fig. 4.5 The convex region in the connected space of 'some'

invariant even when the intersective region forms an(other) intersection with the domain set, that is, with the set of children in our example. That is to say that we can equally faithfully say that some children are naughty children. The new intersection between the set of children and the set of naughty children will return the same focused cognitive space demarcated in Fig. 4.5. This is exactly what the Conservativity Constraint requires. The focused cognitive space, which forms a convex region with the backgrounded cognitive space, is the space that remains invariant over possible (re)-configurations, while the cognitive space *minus the focused cognitive space* does not (necessarily) form a convex region. This indicates that the fulcrum of convexity in the diagrams is a signature of Conservativity. This will have consequences for the behavior of those quantifiers that do not obey the Conservativity Constraint but otherwise respect the Witness Set Constraint.

Likewise, the case of 'no' deserves consideration since the cognitive structuring of quantificational linguistic structures for negative quantifiers such as 'no' will be an addition to the explanatory value of the cognitive description of linguistic structures. This quantifier has something interesting to offer because the quantifier 'no,' unlike other quantifiers, is symmetric in having no directionality in the inclusion/exclusion relations. If we say 'no human is perfect,' the domain set (i.e., the set of humans) and the restriction set (i.e., the set of perfect individuals) cannot have an intersection which contains even a single element. In other words, there is nothing that is shared between the domain set and the restriction set, which apparently gets the cognitively structured space corresponding to the domain set juxtaposed with the cognitively structured space consisting of the restriction set (and/or its complement). It does not matter which set (the domain set or the restriction set) goes first for the intersection. The same holds true for 'some' as well. However, what seems to distinguish 'no' from 'some' is that the foregrounded cognitive space in the case of 'no' will be constituted by the domain set whose intersective value with the restriction set is null, whereas the focused cognitively structured space in the case of 'some' is clearly the intersective region itself. Notice that when we say that no human is perfect, we do not simply mean to say that no one who is perfect is a human. Although this is formally true so long as the set of perfect individuals has no human member, this does not fully reflect the cognitively structured foregrounding/backgrounding relations. This is so because the emphasis is not so much on the non-human quality of perfect individuals in some (possible) world as on the imperfect quality or character of humans. This intuitive component of the cognitive structuring is lost in the symmetric character of intersections. This indicates that the focused or foregrounded cognitively structured space must be configured by the domain set itself with the restriction set (possibly along with its complement) forming the backgrounded cognitive space.

In Fig. 4.6, it is clear that the focused cognitive space constituted by the domain set (i.e., by the set of humans in our example) forms a convex region just like the backgrounded space constituted solely by the restriction set (i.e., by the set of perfect individuals) does. But these two spaces do not in tandem give rise to any convex region, as indicated by the line connected by the points A and B in Fig. 4.6. Further, it may be noted that since the domain set and the restriction set have no intersection, it is trivially true that any further intersection between the domain set on the one hand and the intersection between the domain set and the restriction set on the other will also turn out to be null. That is to say that if it holds that no human is perfect, it is obvious that no human is a human that is perfect. Thus, Conservativity obtains in a trivial manner. This is by virtue of the fact that the focused cognitive space in Fig. 4.6, unlike the focused cognitive space in 'some,'

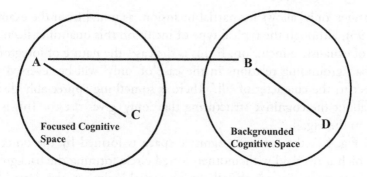

Fig. 4.6 The convex and non-convex regions in 'no'

forms no convex region with the backgrounded cognitive space. Since the focused cognitive space and the backgrounded cognitive space together form no convex region whatever, the absence of convexity in the case of 'no' is akin to the absence of connectedness between the two regions as all convex regions are by default connected but not vice versa (see for details, Gärdenfors 2000). To put it in other words, the absence of convexity between the focused cognitive space and the backgrounded cognitive space returns nothing but the focused cognitive space juxtaposed with the backgrounded cognitive space. This amounts to saying that it is not only that any further intersection between the *domain set* on the one hand and the intersection between the domain set and the restriction set on the other will turn out to be null but also that an intersection between the *restriction set* on the one hand and the intersection between the domain set and the restriction set on the other will also be null. This underscores what the Conservativity Constraint warrants. Thus, convexity (either its presence or its absence) marks Conservativity in cognitively structured spaces of quantificational linguistic structures.

Against this backdrop, one may now wonder how and in what ways the quantifiers 'only' and 'many' differ from these quantifiers in their cognitive structuring so that the linguistic differences can be recast as significant and distinctive differences in the cognitive structuring of 'only' and 'many.' Let's see how to go about it. We are aware that the quantifier 'only' allows for partial inclusion, as noted from the examples (91–93), although the typical type of inclusion this quantifier licenses is that of exhaustive inclusion. If this is the case, the nature of foregrounding/backgrounding relations in the case of 'only' will be reversed with respect to the character of 'all.' There is something appreciably distinctive about the cognitive structuring that 'only' gives rise to. This is evident in the Fig. 4.7.

In Fig. 4.7, the focused cognitive space is formed by the outer circle which is overlaid with another dotted circle forming the background cognitive space. Since 'only' allows for partial inclusion, the dotted lines in the inside circle indicate the possibility of the backgrounded cognitive space being enclosed within, rather than covering, the focused cognitive space. What is interesting here is that it is the backgrounded cognitive space that covers the focused cognitive space in such a way that the focused cognitive space and the backgrounded cognitive space merge into a single unified region, and this unified space forms a convex region. The line connected by points A and C shows this clearly. If we take the example in (91), we immediately recognize that the set of parents of the kids gives rise to the focused cognitive space which is covered by the backgrounded cognitive space constituted by the restriction set, that is, by the set of individuals who roam freely in the backyard of the school—a case of the restriction set being included within the domain

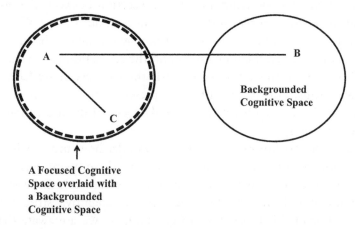

Fig. 4.7 The convex and non-convex regions in 'only'

4 Linguistic Structures as Cognitive Structures 179

set. While the focused cognitive space merged with the backgrounded cognitive space opens up a convex region in the unified cognitive space constituted by the two, the focused cognitive space forms a non-convex region with another backgrounded cognitive space. The line linked by points A and B displays the non-convex region formed by the unified convex region on the one hand and the disconnected backgrounded cognitive space on the other. This new backgrounded cognitive space is way different from the other backgrounded cognitive space fused into the focused cognitive space. In fact, this disconnected backgrounded cognitive space is more primary and central to the meaning of 'only.' Recall that in the example of (91) if it is found that there is a parent of the kid(s) that does not roam freely in the backyard of the school, it does not cause so much harm to the satisfaction condition of the meaning inherent in 'only.' But, if it is found that there is someone who is not a parent of the kid(s) and yet roams freely in the backyard of the school, it provides a strong counter case for the meaning of 'only.' As a matter of fact, the contrastive focus present in 'only' makes sense against this excluded set, that is, the set of non-parents of the kid(s) in (91), for example.

Therefore, the backgrounded cognitive space which is disconnected from the other backgrounded cognitive space fused into the focused cognitive space is more salient with respect to the structuring of foregrounding/backgrounding relations in 'only.' What is strikingly idiosyncratic in the organization of cognitive structuring here is that the focused cognitive space forms a convex region in fusion with the peripherally important backgrounded cognitive space on the one hand, and yet on the other hand forms another non-convex region with another backgrounded cognitive space which is more central to the meaning of 'only.' This reveals that the cognitive structuring involved in 'only' combines the cognitive-structural properties of 'all' and 'no' together. While 'all' gets the focused cognitive space enclosed within, or potentially merged with the whole of, the broader backgrounded cognitive space, the quantifier 'no' gets the focused cognitive space disconnected from the backgrounded cognitive space. The reason why the quantifier 'only' does not go by the Conservativity Constraint becomes clearer now. Since the focused cognitive space and the more primary backgrounded cognitive space are disconnected from each other, any intersection between them, and also between one of them and the previous intersection itself, will return nothing. This may be supposed to establish Conservativity just like it does in the case of 'no,' but this is not possible because Conservativity has been defined with respect to the restriction set that can be individuated independently of the domain set, regardless of whether or not the domain set itself contains its restriction in the form of this hidden backgrounded cognitive space. Besides, quantificational linguistic structures containing 'only' as a quantifier do not have to express this backgrounded cognitive space because it always remains implicit in the presupposed background as a specific restriction on the domain set. And even if it is expressed as in 'Only children but not adults are allowed inside the park,' the set-theoretic relation of exclusion to be defined with respect to the set of adults renders 'only' more complex. That is, the quantifier 'only' is no longer simplex; it turns into a *distinct* complex quantifier (something like 'all but one,' 'many but not Max,' 'more than ten,' etc.). Plus even when this situation obtains, there may still remain a set of non-children in the background which cannot be exhausted by the set of adults since all non-children need not be adults (a table is a non-child but whether it is an adult is out of question). In the face of this difficulty, we may ultimately express the whole excluded set itself as in 'Only children but not non-children are allowed inside the park,' but this serves no demarcating purpose as this set is *already* automatically excluded by the use of the quantifier 'only,' and hence, it is both linguistically and logically redundant.

The disconnectedness between the focused cognitive space and the pivotal backgrounded cognitive space leaves us with just the convex region in Fig. 4.7, while we cannot appropriately articulate the linguistic relation with the disconnected backgrounded cognitive space which is surely more primary than any other backgrounded cognitive space. Thus, as we zoom in on the convex region, it turns out to be the case that the domain set and the restriction set are identical when they are subsets of each other. This naturally equates to a tautology. This exactly chimes with the canonical interpretation of 'only.' From the vantage point of cognitive structuring, the non-convex region formed by the focused cognitive space with the disconnected backgrounded cognitive space contrasts, and goes in conflict, with the convexity in another distinct region formed by the same focused cognitive space along with another backgrounded cognitive space. This incompatibility between convexity on the one hand and non-convexity on the other creates a kind of ambivalence in the cognitive structuring of the quantificational linguistic structures with 'only.' The case of partial inclusion in 'only' is not any good either. One may also note that even when the restriction set is a proper subset of the domain set in the case of partial inclusion licensed by 'only,' this incompatibility remains intact as the focused cognitive space preserves convexity in its own region with the enclosed backgrounded cognitive space.

Now we may look into the cognitive structuring of 'many' in order to understand why it does not obey the Conservativity Constraint. It may be noted that the character of 'many' is different from that of 'only' because 'many' manipulates foregrounding/backgrounding relations in ways dictated by various factors of contextual properties. Although the context-varying character of 'many' may lead one to conclude that 'many' is not at all a kind of quantifier in the same sense in which 'all,' for example, is a quantifier. However, the line of reasoning in the present context would aim to uncover the cognitive structuring hidden beneath the context-varying interpretative property of 'many' in order to project insights into its cognitive character that reveals more about its linguistic variability.

If we look closely at Fig. 4.8, it is remarkably clear that the cognitive spaces involved in the structuring of linguistic structures containing 'many' intertwine with each other in a slightly more complex manner than those of 'only.' This is because 'many' has a context-sensitive property which makes for a degree of sophistication in the cognitive structuring. It needs to be made clear that the cognitive spaces marked with solid-line circles are cognitive spaces corresponding to the two proportions that are evaluated in the inequality in (94). Thus, if we take up the same example which has been taken to illustrate the equation in (94) (i.e., 'Many books in the library are on thermodynamics'), the proportion of the books in the library that are on thermodynamics will constitute the focused cognitive space, and the proportion of the books

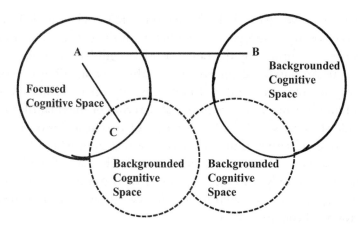

Fig. 4.8 The convex and non-convex regions in 'many'

in the library that are not on thermodynamics will constitute the most primary backgrounded cognitive space. Hence, they are marked with solid-line circles which represent the two contrasting cognitive spaces which do not form a convex region, as indicated by the line connected by points A and B. But the focused cognitive space can form a convex region with (parts of) another backgrounded cognitive space which is not primary or at least less central than the one with which it forms no convex region. As a matter of fact, it is the shared space between the focused cognitive space and this new backgrounded cognitive space that forms a convex region with the rest of the focused cognitive space. This is marked by the line connected by points A and C. But it may be observed that the circle representing this backgrounded cognitive space is in dotted lines. This is shown to emphasize the point that this is a hidden or contextually implicit cognitive space that may or may not emerge in certain delimited circumstances. This backgrounded cognitive space may be individuated by the proportion of all those entities that are on thermodynamics, to take the case of our present example. Since all proportions, when calculated in a certain context, are to output a number of entities, there can be an intersection between two proportions. For instance, the proportion of the books in the library that are on thermodynamics will yield a certain number of books when the required values are provided. If so, the proportion of all those entities that are on thermodynamics will produce a number of books, articles, essays, etc. some of which may also be found in the number of books output by the calculation of the proportion of the books in the library that are on thermodynamics. But contextual considerations do not just stop here. We can also have another contextually implicit cognitive space that specifies the proportion of all entities that are not on thermodynamics, and this is marked by the other circle in dotted lines onto which the first contextually implicit cognitive space is superimposed.

The reason why 'many' fails to comply with the Conservativity Constraint is that 'many,' just like 'only,' also gives rise to an incompatibility between convexity and non-convexity in its cognitive structuring of spaces. On the one hand, it forms a convex region with a backgrounded cognitive space when the shared space between the focused cognitive space and this backgrounded cognitive space forms a convex region with the rest of the focused cognitive space, but on the other hand it forms a non-convex region with the most central backgrounded cognitive space—the cognitive space against which the very proportion that forms the centrally focused cognitive space is defined. Now the patterns of commonality bringing quantifiers such as 'all,' 'some,' 'no,' etc. under the unifying ambit of the Conservativity Constraint have been (re)expressed in terms of their underlying cognitive structuring. It turns out that the subtle patterns of organization of cognitive spaces in terms of convexity/non-convexity shaping and molding quantificational linguistic structures are largely responsible for the linguistic effects that are observed. One of the significant corollaries emerging from this exercise is that all the quantifiers examined so far obey the Witness Set Constraint, precisely because the cognitive structuring of each of the quantifiers looked at above exhibits a convex region within the focused cognitive space, and this remains invariant over all possible transformations and alterations conceivably performed. It may be noted that the Witness Set Constraint will always be obeyed since any subspace corresponding to a subset of the relevant domain set within the focused cognitive space must also form a convex region. This ensures that it is not simply the focused cognitive space itself but also a subspace within it that can, by virtue of preserving convexity within, pave the way for the consolidation of the Witness Set Constraint. This may be believed to furnish a deeper and more fundamental grounding of the intricate behaviors of quantificational linguistic structures. A few more linguistic phenomena need to be scrutinized with the same goal in mind. This is what we shall turn to now.

4.1.3 Complex Predicates

If there is an essential sense in which the cognitive structuring of events in linguistic structures comes out in a strikingly organized manner, it must be complex predicates in natural language. Predicates in natural language are like functions that take designated arguments as inputs and tell us whether the inputs belong to sets denoted by the predicates. So, for instance, if 'amazing' is a predicate, it can take 'Titanic,' for example, as an argument and return the answer as to its membership in the set of things that are amazing. It is not just adjectives—nouns, verbs, prepositions all can act as predicates when they take arguments. Thus, a noun such as 'fear' can take as an argument something that someone has a fear of, say, spiders. But this cannot be taken to mean that all nouns are 'fear'-like predicates—when used in a linguistic structure the noun 'human,' unlike nouns like 'fear,' 'envy,' 'security,' 'link,' or 'care,' does not take any argument in the same way as 'fear' does in the relevant syntactic position (we cannot thus say 'He is a human *of* something/someone'). On the other hand, logically speaking, the noun 'human' when used as a noun does take another argument when we say 'Hem is a human' because the noun 'Hem' belongs to the set of entities denoted by the predicate 'human.' This can be represented as 'HUMAN(Hem).' This makes it obvious that predicates can take more than one argument in specific syntactic positions whose roles usually differ from each other. Thus, a noun like 'fear' houses two arguments—one is what can be designated as fear and the other is what a certain fear is a fear of. The only existing argument in 'human' is generally considered to be *external* because it appears in a syntactically distant position from the predicate 'human,' whereas the second argument in 'fear' is taken to be *internal* since it appears close to the predicate, or simply hangs on to it.

4 Linguistic Structures as Cognitive Structures

Therefore, predicates are logically structured entities that require arguments for the satisfaction of what they denote. It is this logical sense of predicates that we may appeal to when talking about (complex) predicates.

Complex predicates are those entities that are formed by combining two or more predicates in natural language. Since complex predicates can be formed by way of combinations of two or more nouns or prepositions or even adjectives, in the present context we shall restrict ourselves to complex predicates formed by two or more verbs. The reason for choosing complex predicates formed by verbal predicates is that verbs constitute the pivot of the event structure of linguistic structures in any type of construction. This indicates that the structure of events encoded in linguistic structures revolves around the form verbal predicates take. When verbal predicates are simplex the nature of events is also plainly uncomplicated, but when verbal predicates are complex the event structures encoded in complex predicates turn out to have cognitive configurations which are not very straightforward. The cognitive constitution of complex predicates can not only tell us a lot about the ways of conceptualizing events but also reveal why certain linguistic patterns in complex predicates exist while certain others do not. This means that the cognitive constitution of complex predicates may also be expected to project insights into the idiosyncrasies of linguistic structures involving complex predicates. Complex predicates are found aplenty in West African, Australian, and South Asian languages. Hence, we shall look at constructions from these languages in order to find more about their cognitive constitution.

Complex predicates formed by a combination of verbs are generally of two types. One type is composed by the combination of one content-unchanging verb (its original meaning remains intact) and one light verb whose original meaning is lost, and the other type is constructed by having a number of verbs (two or more) strung out along an ordered sequence, although one among these verbs can be a light verb. The latter type is generally called *serial verb constructions*. The following sentences in (99) and (100) are typical serial verb constructions from respectively Barai and Yimas, two Papua New Guinea languages.

(99) fu burede ije sime **abe ufu**
      he bread DEF knife take cut
      'He cut the bread with the knife.'   (Foley and Olson 1985: 44)
(100) narm   pu-**tpul-kamprak**-r-akn
      skin-SING 3PLU hit-break- PERF -SING
      'They hit and broke his skin.'   (Foley 2010)

And the following sentence is from Dagaare, a West African language.

(101) o   da   mong la   saao **de bing bare ko** ma
      3.SING PAST stir FACT food take put leave give me
      'S/he made food and left it there for me.' (Bodomo 1997: 15)

[DEF = definiteness marker, PLU = plural, PERF = perfective aspect marker, FACT = factivity marker].

As can be recognized above, the verbal predicates that occur in a sequence are marked in bold. While the examples in (99–100) contain two predicates, the sentence in (101) has three predicates in a sequence. In this connection, it is also worthwhile to note that the events specified by the verbal predicates *can* be interpreted in terms of the sequence in which the predicates appear in the structure. Thus, for example, in the sentence (99) one could say that the male individual took a knife and used it as an instrument to cut the bread, and in (100) the hitting event (causally) led to the breaking event as is evident in the serial order of the predicates on the surface. The example in (101), however, differs from (99) and (100) in that the predicate 'ko' (meaning 'give') does not, strictly speaking, contribute its event structure to the overall interpretation of the sequence of events. That is to say that the sentence does not mean that the person made food and left it and gave it to the speaker. Rather, the predicate 'ko' seems to introduce a new argument which is not licensed by either 'bare' (meaning 'leave') or 'bing' and which turns out to be the *recipient* of the (internal) argument of 'bare'/'bing' (i.e., 'saao' meaning 'food')—a point also noted in Baker and Harvey (2010). Nonetheless, it is not hard to re-interpret this sentence as meaning that the person made food, took it, and left it in order to give it to the speaker. The serial interpretation of the order of the verbal predicates is then preserved. But, regardless of how this is interpreted, what remains crucial is that the sequence of predicates in serial verb constructions matters, even when the exact position of a certain verbal predicate in the serial order may not be immediately indicative of its role *as an event* in the overall meaning of a linguistic structure. This carries important consequences for what will be later said about the cognitive structuring of events revealed by complex predicates.

If serial verb constructions specify the sequencing of events assembled under a single linguistic structure, complex predicates formed by two verbal predicates convey something significant about the *micro-structural* properties of events. To understand what these properties are, let's look at some representative constructions of this kind.

(102) **birli**   nga-Ø-**ganji**
       go.in 1 SING -3 SING- take
       'I put it in(side).'

(103) **rang** ng- **anyi**   Ø-manuga
       hit 1 SING/3 SING-take rock
       'I hit a rock.'   (Baker and Harvey 2010: 16 & 14)

The sentences above are from the Australian language Marra, whereas the following two sentences are respectively from the South Asian languages Hindi/Urdu and Bengali.

(104) us-ne   khana **kha**   **li**-ya
       he/she-ERG food   eat-NONF   take-PERF
       'He/she has eaten food.'

(105) ami pata-ti   **chhire**   **phel**-l-am
       I   leaf-DEF tear-NONF   throw-PAST-1 SING
       'I tore the leaf (into pieces).'

[ERG = ergative case marker, NONF = nonfinite, LOC = locative marker, PERF = perfective].

The two verbs involved the formation of complex predicates in (102–105) above have been marked in bold, as has been done for serial verb constructions before. In each case, the verb that appears second in the sequence is the verb that has lost its original semantic content—it is the light verb. In other words, the first verb in the sequence is the verb that has retained its semantic content. But this need not be the universal case. What is important here is that the two verbs involved contribute to the overall meaning of the sentences by specifying certain properties of the structure of events encoded (see also Butt 1995). For instance, in (102) the verb 'ganji' contributes an internal argument that can be hooked onto the verb 'birli' which does not take any internal argument of its own. So the light verb 'ganji' actually relates the relevant event to a specific participant in the event (which is the item moved/put aside), which would not have come about with only 'birli' because 'birli' does not encode in its argument structure the participant item moved. But in (103) the situation is somewhat different. Both the verbs in being transitive verbs can take internal arguments, which is what underpins their commonality, but there is still something special that 'nganyi' adds to 'rang.' Notice that the verb 'nganyi' (meaning 'take') incorporates both the external argument (1 SING) and the internal argument (3 SING) in its meaning, especially when it is used independently as a verb. Since 'rang' (meaning 'hit') also requires an internal argument (the object participant which is hit), the internal argument of 'nganyi' can be unified with that of 'rang.' This part seems clear. But, more importantly, when we home in on the relation between the overall event and external argument, that is, the agent participant (which is the speaker herself), it becomes evident that the event encoded in 'nganyi' specifies a movement caused by the agent (its external argument) and 'rang' specifies a manner of movement—a movement *with force* toward something. Therefore, 'nganyi' contributes the component of movement caused *by an agent* to a manner of movement. This results in a movement caused by an agent with force onto the object participant, that is, the rock.

On the other hand, the examples in (104–105) are quite different from those in (102–103). The verb 'liya' (meaning 'take') in (104) adds something on a distinct dimension to the verb form 'kha.' Without doubt, the event of eating is modified in subtle ways, and the difference can be discerned by comparing it with its alternative version minus the light verb 'liya' in (104') provided below.

(104')   us-ne   khana   kha-ya
         he/she-ERG   food   eat-PERF
         'He/she has eaten food.'

4 Linguistic Structures as Cognitive Structures 189

While (104') is also perfective, that is, the event of eating can be said to have been completed, the result of the completion of the event of eating stays or *sort of* 'lingers on' in (104'). One cannot usually, *at the exact endpoint* of the eating event, say (104') to mean that the person in the context has finished eating, while (104) would be perfectly natural in such a scenario. In a scenario in which (104') is used as a question with a high intonation toward the end, the answer cannot contain 'kha liya' if, for example, the person concerned ate food one or two hours before the time point at which the question is asked. This is so because the immediacy associated with 'kha liya' is lost. An answer containing 'khaya' would be appropriate. That is to say that the light verb 'liya' in (104) adds 'boundedness' not only to the main event but also to the sub-event that follows the main event of (only) eating expressed by 'kha.' Likewise, the light verb 'phellam' in (105) also adds a sort of 'boundedness,' not to any sub-event, but to the main event of throwing itself. Just like (104'), the alternative version of (105), given below as (105'), does not convey the same thing as (105).

(105') ami pata-ti chhir-l-am
I leaf-DEF tear-PAST-1SING
'I tore the leaf (into pieces).'

The evidence for the claim that (105') conveys a bounded event is to be found in the sentences (105"–105"') below.

(105") ami pata-ti koyek minit dhore chhir-l-am
I leaf-DEF several minute for tear-PAST-1SING
'I tore the leaf (into pieces) for several minutes.'
(105"') *ami pata-ti koyek minit dhore **chhire phel-l-am**
I leaf-DEF several minute for tear-NONF throw-PAST-1 SING
'I tore the leaf (into pieces) for several minutes.'

Since the light verb 'phellam' adds a sense of instantaneousness to the main event of tearing the leaf, a temporally stretched event of tearing the leaf is utterly incompatible with the use of the light verb 'phellam,' as shown in (107). In sum, light verbs in complex predicates formed by two verbs not only relate events to event participants in certain non-trivial ways (as revealed through (102–103)) but also modify and re-shape the structure of events encoded in the main verbs (as shown in (104–105)).

Thus far the discussion has conveyed the impression that light verbs in virtue of (re)structuring events or their relation to event participants do not at all contribute their original semantic contents to the event complex. This is not so. Light verbs can, of course, retain their original meaning in complex predicates, thereby belying the meaning of the term 'light verb,' albeit in a restricted context. A sentence from another Australian language Wagiman demonstrates this well.

(106) bak   Ø-linyi-ng   lari
       break 3 SING-fall-PAST   arm
       'He fell and broke his arm/He broke his arm in falling.'

(Wilson 1999: 104,105; also produced in Baker and Harvey (2010))

In such cases, it appears that light verbs may also do something quite different from what they usually do in the kind of verbal complex predicates shown above. The light verb here is 'linyi' ('fall') which specifies the causal process along a path that led to the event of the person's breaking his arm. Clearly, the light verb here re-shapes the structure of the event of one's (accidentally) breaking own arm, but it does so not by contributing boundedness to the main event of the breaking of the arm. Rather, a portion of the main event of the breaking of the arm—the initial portion, in fact—is singled out and then related to the (sub-)event that initiated or caused the main event. In short, even here the relevant light verb restructures the event encoded in one of the verbs participating in the complex predicate formation.

Now one of the most significant questions that may preoccupy us here is whether there could be a good way of marking a principled cognitively structured distinction between verbal complex predicates and serial verb constructions. Note that here one must also take into account the fact that the common structural theme binding them both is the formation of a type of complex predicate by combining two or more verbs. That is, the relevant distinction between verbal complex predicates and serial verb constructions has to be juxtaposed with the commonality underpinning them both, in order that both facets can be brought under the same rubric of explanation. One proposal that takes has attempted to approach this issue is to be found in Baker and Harvey (2010) who have postulated that verbal complex predicates that are formed by combining two verbs characterize singular or 'simplex' events. The eventive representations of such simplex events come from two different verbal predicates when the conceptual structures from the two verbs become fused, or simply merge, into a single whole. Serial verb constructions, on their view, are conceptually distinct from two-verb complex predicates because they instantiate more complex events—events that cannot be considered atomic or indivisible. To put it in a different way, serial verb constructions are distinguished by their incorporation, not of a single event, but rather of a number of events put together in a single construction. But this proposal has been contested by Foley (2010) who argues that serial verb constructions can implement simplex events on the one hand and, on the other hand, often give rise to intricate patterns of combination of events when they denote complex events. An example of a simplex event encoded by a serial verb construction is from the Papua New Guinea language Numbami.

(107) kolapa i-lapa bola uni
 boy 3SING -hit pig dead
 'The boy killed the pig.' (Foley 2010: 85)

Below is an example of a complex event in which two events out of three in a serial verb construction are closely knit together while another single event stands somewhat isolated. This example is from Watam, another Papua New Guinea language.

(108) yor i aŋgi-mbe pika-(*mbe) irik
 egg a get- IRR throw- IRR go.down
 'Get an egg and throw it down!' (Foley 2010: 101)
 [IRR = irrealis marker]

The event denoted by 'aŋgi' (meaning 'get') stands somewhat detached from the complex event formed by the verbs 'pika' ('throw') and 'irik' ('go down'). This is exactly what is indicated by the ungrammaticality of the irrealis marking suffix '-mbe' between 'pika' and 'irik,' while the presence of this marker between 'aŋgi' and 'pika' signifies the weaker integration of the initiating event into the entire event complex formed by the three verbs in a sequence. This indicates that the sequence of events encoded in serial verb constructions need not have the individual events uniformly strung out along a line. Rather, some events may form certain combinations based on their simultaneity, spatial closeness or natural order, whereas some others remain separated from such combinations.

In the face of such a dilemma, one may now wonder whether there is really any conceptual distinction to be made between verbal complex predicates and serial verb constructions since, after all, they are both complex predicates but of different types. Note that this issue arises even when one does not adopt an analysis in terms of the apparatus of Lexical Conceptual Semantics (see Jackendoff 1990, 2002; Levin and Hovav 2005). This is so because one has to supply an account of the nature of event combinations or of the conjoined events in verbal complex predicates and serial verb constructions on independent grounds. Clearly, there are linguistic differences between the two types, as has been shown above, but if the linguistic differences fall straightforwardly out of the difference in their cognitive structuring, the account of the linguistic differences will become not only enriched but also more meaningful. With this in mind, let's see how to go about this.

First of all, we may concentrate on the character of the microstructural properties of events that verbal complex predicates encode, represent, manipulate, and re-shape. It has been noted above that verbal complex predicates formed by two verbs sometimes relate events to event participants in certain non-trivial ways and also modify or re-shape the structure of events encoded in the main verbs by individuating and manipulating certain parts of events. This appears to capture some of the *intra-eventive* properties in the sense that one needs to look *inside* an event in order to relate events to event participants or even to represent and re-shape parts of events. These intra-eventive properties are the exact correlates of micro-structural properties of events—properties that have to be discovered within events. We have to appeal to these properties if we want to make headway toward understanding the difference between verbal complex predicates formed by two verbs and serial verb constructions. If the question of making out the difference between verbal complex predicates and serial verb constructions revolves around the question of whether one of these two types of constructions encodes simplex events while the other does not, we certainly cannot get closer to the appropriate explanation. Rather, we have to look beyond the question of whether the relevant events encoded by either of the two types of constructions are simple or complex. That is, regardless of whether the events encoded by these constructions are simple or complex, there has to be something irreducibly distinctive about verbal complex predicates that isolates them from serial verb constructions.

Note that constructions with verbal complex predicates say something about *intra-eventive* properties. This indicates that when one requires the projection of a fine-grained view of events, constructions with verbal complex predicates would serve to express this in a more compact manner. But when a relation between events needs to be stated, serial verb constructions will be handy enough. At this stage, one may feel that this is just another way of recasting the simplex event/complex event difference between verbal complex predicates (formed by two verbs) and serial verb constructions in different terms. But, in fact, this is not so. Even when the relevant events encoded by serial verb constructions are simplex events, as in (107), it can be shown that the constituent events and/or states are linked or chained to each other in ways that cannot be realized within constructions with verbal complex predicates. Similarly, on the other hand, even when the events encoded by constructions with verbal complex predicates seem to be complex, it is not hard to detect the projection of micro-structural properties of events inside such constructions. Before we get down to the rationale for thinking this way, let's dig a bit deeper into these constructions to find out the most essential properties that uncover their cognitive structuring.

When complex predicates are formed by combining two verbs, the light verb usually contributes something to the overall meaning of the construction. In fact, it contributes to the interpretation of the event structure complex predicates constitute. It does so by relating the event encoded *either* to certain event participants (agents, themes or patients, i.e., entities on which agents act and are thereby affected by their actions) *or* to certain instants/times within the main event or within some other preceding or following sub-event. In order to zoom in on the event-internal properties, light verbs sort of *magnify* certain relations that obtain between the main events and certain relevant event participants, as in the examples (102–103). Additionally, light verbs can also foreground significant relations between the main events and certain instants/times within the main events or within some other preceding or following sub-event—something that applies well to cases like (104–105). Now it may not be immediately clear how complex predicates, in virtue of containing light verbs, relate main events to instants/times located within the main events or within some other preceding or following sub-event.

It needs to be understood that when the event of eating in the example of (104) above is said to carry a sense of immediate completion due to the presence of the light verb therein, the light verb re-shapes the *aspectual* profile of the event rather than the tense itself. Re-structuring aspectual profiles of verbs is tantamount to altering the internal structuring of instants/times that constitute an event. Thus, for example, a process component can be added to, or eliminated from, an event when an aspectual alteration takes place. A simple example can illustrate this well. The verb 'run' is a predicate that specifies an *activity* which incorporates a *process* without any natural bound or endpoint, as in 'She ran in the garden.' Now if we say 'She ran to the garden,' the prepositional phrase 'to the garden' imposes a bound on the process involved in the activity of running, thereby placing constraints on the amount of the process to be found in the event of running. Similarly, when the light verb re-shapes the aspectual profile of the event in (104), what it actually does is relate the main event of eating to the instants/times located *at the endpoint* of the event of eating. By doing this, the light verb suppresses the process component of eating and highlights the final instants/times of the event of eating and carries it all over to the immediately following sub-event that bears the effect of eating. Likewise, in the example (105) the light verb delimits the process of tearing and then relates the main event of tearing to the exact instants/times that mark the endpoint of the main event. This is what is meant when complex predicates are said to relate main events to instants/times located within the main events or within some other preceding or following sub-event.

Given this analysis, one may now wonder how this extends to the example in (106) and other similar cases found in languages. On the one hand, the light verb here does not behave like a verb bleached of its usual semantic contents, and on the other hand, it seems to add another (sub)event to the main event of (accidentally) breaking one's arm, thereby rendering the entire event complex. Independent of whether the overall event is actually complex or simple, what happens even here is that the main event of one's (accidentally) breaking one's own arm is related to the very instants/times situated at the endpoint of the falling event. It is not so much the entire process involved in falling as the final moments or instants of the falling event that matter for the relation between the main event and the preceding sub-event. This point has also been noted in Baker and Harvey (2010). If this is the case, one need not really bother about the exact nature of the preceding sub-event that led to the event of one's (accidentally) breaking one's own arm. It may be that the preceding sub-event played a *causal* role in leading to the main event, or that it simply initiated the main event of breaking without being a direct cause of it, or even that it specifies the manner in which the main event took place. This question is immaterial. That is because when serial verb constructions are found to express a relation between a causal event and the caused sub-event/state, as in (107), one may leap to the conclusion that serial verb constructions and constructions with verbal complex predicates (formed by two verbs) are, after all, alike since both can encode a relation between a causal event and the caused sub-event/state. As a matter of fact, this issue is the source of the contention revolving around the question of whether events encoded in serial verb constructions are complex or simple. There is no denying the fact that serial verb constructions do encode relations between a causal event and the caused event/state concerned, as Foley (2010) points out that the Papua New Guinea language Watam expresses 'kill' in various ways in serial verb constructions by combining a verb specifying the manner of killing with the verb of dying (making someone die by spearing/shooting/hitting, etc., for example).

However, it is worthwhile to note that even though the causal or initiating event is specified in both serial verb constructions and constructions with verbal complex predicates and the main event is related to it, the nature of the *relatum* differs in constructions with verbal complex predicates. The main event is related *not* to the whole sub-event that precedes and causes the main event *but rather* only to those parts of the initiating/causal sub-event that are immediately adjacent to the onset of the main event. The required evidence for it comes from (107) itself. As Foley (2010) states that the verb '-lapa' in (107) 'denotes any act of killing, whether by clubbing, stabbing, or shooting, and is not restricted to causing events with a specific manner of action,' this clearly indicates that it is the entire event of some unspecified manner of action that is most relevant, but not the specific structural components of a particular manner of action event. Similar considerations apply to the case of 'kill' in Watam since in each case of a given manner of action event the final instants of that event are irrelevant. Had this been not the case, the structural composition of events in such examples would suggest that the resulting state of one' dying is *always* at the exact endpoint of the manner of action event, no matter what manner of action event causes the resulting state of dying. In such a scenario, the difference in the process involved in each particular manner of action event would be muddied or erased. Here all that matters is that the manner of action event causes the death of the person to whom the causal event applies. This is because one may stab someone without the person stabbed becoming immediately dead, but one may hit someone who dies immediately as a result of the hitting. The case of 'kill' in Watam is, thus, a case of specifying different means (events of different manners of action) of achieving the same end (making someone die), but not of specifying different kinds of final instants/times contemporaneous with the same effect (dying, in this case). While this holds true for serial verb constructions, whatever has been said about the character of verbal complex predicates remains valid even when the sub-event is not a causal event in a given relation between the main event and certain instants/times located within some sub-event.

(109) durdut bula-ndi
      run   leave-PAST
      'She ran away from him/She ran away and left him.'

   (Wilson 1999: 154–155; also produced in Baker and Harvey (2010))

Here the main event of running is related not to the process of leaving per se but rather to the exact point/instant at which the leaving action is completed. Again, this is independent of whether one considers the whole event encoded in (109) complex or simple.

With this in place, we are now poised to go into the exploration of the event structure serial verb constructions give rise to. The central point of the present proposal is that if verbal complex predicates project relations between events and specific event participants or certain instants/times located within some (sub-)event, serial verb constructions give rise to relations between events. Relations between events provide ways of linking one event to another and then to other events. Regardless of whether the whole complex of conjoined events is itself a simple event or a complex one, what matters here is that the required relation obtains between events, *not between* an event and its micro-structural properties. Whether some event, especially in the case of events encoded in serial verb constructions, is simple or complex does not, at least in a straightforward manner, hinge on the nature of event itself. Rather, it often depends on how the events linked are assembled by the linguistic structures in such constructions. Thus, for instance, the case of causing a pig to die by hitting it in the Numbami example (107) could be conceived of as a single event since the exact manner of action event is immaterial for the resulting state. But there is nothing that can also prevent it from being interpreted as a complex event if the act of causing some living entity to remain not alive any more is *conceptually* segregated from the way that act is accomplished. This is so because a given manner of action that can *potentially* cause the death of a living entity can be executed completely independently of the intention to kill that living entity. One can, of course, club someone having no purpose of killing that person. Conversely, one can also entertain the intention to kill someone without thinking up any way of realizing that intention.

Therefore, event-to-event relations encoded by serial verb constructions must be distinguished from event-to-event participant/instant relations encoded in verbal complex predicates formed by two verbs. Now it is necessary to emphasize that event-to-event relations encoded by serial verb constructions can themselves be of various types and patterns. This enables serial verb constructions to encode either simultaneous (or almost simultaneous) events in a sequence (as in (100–101) above) or discontinuously integrated event sequences (as in (108)). What suffices for event-to-event relations to be recognized as such is for them to relate events to other events but not (as a general principle) penetrate into events *only* in order to bring certain internal structural parts of events into correspondence with events. But from the examples (99) and (101) above, it may seem that serial verb constructions do actually get inside events and then relate certain internal structural parts of events to other events by way of relating events to event participants. Thus, one may think that the object argument of the verb 'abe' in (99) is brought out and then the event of cutting the bread is related to the object argument of 'abe' (which is the knife). Likewise, one may assume that the *recipient* argument of the verb 'ko' in (101) is brought out and then the event of leaving the food is related to this argument.

However, careful scrutiny reveals that this is not quite right. If we say that the event of cutting the bread is related to the object argument of 'abe' in (99) by way of going into the event of taking encoded by 'abe,' it appears that the event of cutting the bread is related not to any event participant of its own (because cutting as an event notionally requires only two participants—the agent who cuts and the thing cut) but rather to an event participant involved in some other event. But then, the main event of cutting the bread is not, strictly speaking, directly or immediately related to internal parts of an event per se. So this is certainly not like getting inside an event and then bringing out its structural parts *only* in order to relate the whole event itself to some of these parts. That this is the case is also evidenced by the fact that the verb 'abe' retains its predication in the sense that one could say that the male person took the knife and cut the bread with it. What happens here is that the main event of cutting the bread is related to another event, which is the event of taking, and then, the *theme* participant (the item taken) of the event of taking is mapped onto the main event of cutting the bread by means of *event sharing*. Event sharing is what is generally found in most serial verb constructions—the case in (100), for instance, shows that the hitting event and the breaking event are both shared with the *patient* participant (i.e., the skin of some male person). It is only by virtue of the property of sharing that an event can be mapped onto the event participant of another event to which the former event gets linked. That is to say that the mapping of the internal parts of an event onto some event in serial verb constructions is just a by-product of event-to-event relations, but not the other way round. The converse holds true for verbal complex predicates in which it is only by virtue of relating events to *intra-event* properties that verbal complex predicates end up linking one event to another. So this implies that what serial verb constructions instantiate is the converse of what verbal complex predicates (formed by two verbs) do. This needs to be spelt out in more precise terms.

Given the characterization of the kinds of event structure serial verb constructions and verbal complex predicates independently go on to constitute, we may now state in more precise terms the forms of the cognitive structuring they permit. This can help spell out not only what underlies their linguistically different behaviors but also how they turn out to have common characteristics in certain respects. Let's now make a distinction between two types of cognitive structuring: *intra-event mapping* and *event linking*.

(110) $E \dashrightarrow S(\dashrightarrow E) = R(E, S)\ \&\ R'(S, E)$ Intra-Event Mapping
$E \dashrightarrow E(\dashrightarrow S) = R''(E, E)\ \&\ R(E, S)$ Event Linking

[E = event, S = some intra-event structure: an event participant or set of instants, and $R, R,' R''$ are different *relations* that relate E, S to one another].

Now what the formulation in (110) states is that in the case of intra-event mapping an event is mapped onto (or simply related to) some event participant or onto a set of instants which are part of some event. This characterizes verbal complex predicates (formed by two verbs) in a basic way. On the other hand, in event linking an event is mapped onto (or related to) another event. This marks serial verb constructions. This mapping can be performed more than once if the number of events is greater than two in a given serial verb construction. But there is another mapping present in the formulation that is more relevant from our perspective. Notice that S is further mapped onto E in intra-event mapping, and this mapping is placed within braces. Likewise, in event linking E is further mapped onto S while remaining enclosed within braces. The braces indicate that this mapping is *not* the main function of each type of cognitive structuring; rather, it is an epiphenomenon or a by-product of the primary mapping executed in each type of cognitive structuring. More interestingly, it may be noted that if we start with the initial mapping in intra-event mapping and then reach the last mapping placed within braces, we can start all the way from there and create a new mapping from E back to E at the beginning and then go from E to S. This way we derive the mapping involved in event linking from intra-event mapping. This completes a circle as represented in Fig. 4.9.

But this seems to imply that we can always derive event linking from intra-event mapping. This is not so. It is possible to derive the latter from the former, as shown in Fig. 4.10.

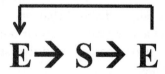

Fig. 4.9 The circular mapping path from intra-event mapping to event linking

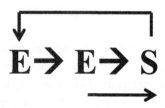

Fig. 4.10 The circular mapping path from intra-event mapping to event linking

The lower arrow in Fig. 4.10 indicates that we have to start in this case from the second E and then map this E onto S and then go in a circular path by mapping S further onto the initial E. This circular trajectory, regardless of the direction we take, represents the idea that intra-event mapping is the *obverse* of event linking and vice versa. At this juncture, it should be compellingly clear that serial verb constructions and verbal complex predicates, distinct though they are, meet at the same point but by travelling through different routes. This accounts for their similarities in terms that appeal to the way they encode event-structural properties. At the same time, this also explains why they behave differently after all. That is crucially because the basic or primary mapping in each case is distinct, even though both these two types of complex predicates arrive at the same point at the end of their individual trajectories. This shows that intra-event mapping as a type of cognitive structuring is a fine-grained mechanism of *event scanning* or event analysis, but this does not in itself preclude the integration of events or event sequences via intra-event mapping. The cognitive plausibility of this can be expounded this way. Whenever events are individually scanned, dissected, and so analyzed, the event structure of a given event has to be assessed as a whole for the analysis of the internal structural parts, and so the whole needs to be juxtaposed with the parts. This is more so when more than one event is involved. Similarly, when events are linked to one another, some internal part(s) will always be shared among the events so linked. In fact, it is the agent that is more often than not shared among the events that are linked to one another. Therefore, it is hardly surprising that intra-event mapping and event linking are complementary cognitive mechanisms that represent, encode, and manipulate event structures in terms of which we linguistically express and conceive states of affairs, situations, scenarios or states, and also their relations to their constitutive parts in the real world.

Moreover, since some internal parts of events may be more salient than certain others in the juxtaposition of whole events with parts, this could be responsible for a wider variation in the selection of certain forms that express various finer subtle distinctions in event-internal properties in verbal complex predicates formed by two verbs. Thus, there could be many-to-many mappings between specific light verbs and the event-internal parts highlighted and related to whole events. This is prevalent in most South Asian languages (see also Butt 1995). While this can also lead to idiosyncrasies in the formation of complex predicates with two verbs, this does tell us something about the cognitive constitution of complex predicates in general. While intra-event relations via the intra-event mapping spotlight a *vertical* projection of the structure of events from a multi-dimensional map of points constituting the internal parts of events, inter-event relations via event-linking project the *horizontal* dimension of event structures whereby events are assimilated, sequenced, compared and grouped into structurally meaningful units.

4.1.4 Word Order

Complex predicates furnish entry into the cognitive constitution of event structures expressed in natural language, whereas the nature of the cognitive representation of a series of distinct items of various categories is revealed by word order in natural language. Word order in natural language has a diversified manifestation, and different natural languages on this planet pick up different kinds of word order systems. Some languages have a highly rigid system of word order, while some others are relatively more flexible in their word order. Because word order varies across languages and different languages have different degrees of rigidity (or conversely, flexibility) in expressing what comes first and what comes next, one may wonder how word order may have something to do with mental representations. But a little reflection tells us something quite different. First of all, word order is all about the order in which words or word categories appear in sentence structures. And if so, a specific order of word categories has to be somehow entrenched in the mind, and other possible comparable word sequences will have to be eliminated from mental tracking against the specific one which is valid in the language. This feat cannot be achieved at all unless there is already in place a suitable way of coding (and also decoding) sequence positions of words or word categories. The special dilemma with natural languages is that this is not as simple as it looks like on a cursory observation. On the one hand, no language in the world has one and only one word order sequence valid throughout the entire language, and on the other hand, different languages display different degrees of flexibility in expressing word sequences. Second, even when a particular sequence of words or word categories is not mentally tracked (either online or in the consolidated associations of the long-term memory), the functional or grammatical roles of words have to be extracted and so internalized from word combinations realized in linguistic structures. That this is not a trivial thing is cross-linguistically evidenced by the presence of linguistic marking on verbs and nouns signaling who did what to whom.

Word order is one of the most easily observable aspects of syntactic structure as it can be discerned right on the surface of linguistic structures. Although word order is visible on the surface, what it has in store for the exploration into its cognitive structuring may not be immediately clear. That is, in part, because of regulations imposed on word order by strict grammatical functions and sometimes due to various pragmatic/discourse functions that constrain possible word order variation within and also across languages. And it is not necessary that the requirements made by strict grammatical functions and the constraints enforced by various pragmatic/discourse functions must be in tune with one another. English, for example, is a language in which word order is genuinely rigid, although we often observe a certain amount of leeway in syntactic structures. The following sentences are just some of the representative examples that illustrate both these aspects of English word order.

(111) These children need a break in June.
(112) We have given them a lot of leeway to make their own way.
(113) Never was I so enthralled by a movie.
(114) This book, I hate so much.

While the sentences in (111–112) exemplify the canonical Subject-Verb-Object word order, the sentences in (113–114) exhibit slight deviations from the canonical word order as the auxiliary verb is placed before the subject in (113) while the object has come before the subject in (114). This may not look like a big deal. But this has consequences for the system of grammar. On the one hand, if the word order system in English is relatively rigid, any amount of flexibility is indicative of pressures employed by other factors such as pragmatic/discourse functions within the language. This is indeed so. The auxiliary inversion with respect to the subject in (113) takes place owing to the fronting of the *focused* negation-marking word 'never,' whereas in (114) the object is fronted because it is given a greater amount of emphasis. In a nutshell, any deviations from the prototypical word order are almost always governed by considerations that do not jell well with those constraints that enforce a rigid word order. In this case, these considerations have something to do with pragmatic/discourse functions since marking something as focused (or even de-focused, for that matter) is not a flawless grammatical function in English (being a subject or object, for example). But this does not, of course, imply that discourse functions cannot have the character of grammatical functions. That this is possible is illustrated by Hungarian.

(115)
a. Janos [IMRET mutatta be Zsuszanak]
   John Imre.ACC introduced prev Susan.DAT
   'John introduced to Susan IMRE.'
b. Zsuzsanak [JANOS mutatta be Imret]
   Susan.DAT John introduced prev Imre.ACC
   'Susan was introduced Imre by JOHN.'
c. Imret [ZSUZSANAK mutatta be Janos]
   'Imre was introduced to SUSAN by John.'   (K. É. Kiss 1998: 682)
   [DAT = dative case marker]

What is interesting in (115) is that in each of the three cases one of the event participants of the introducing event is fronted and then assumes the role of the referent about which the sentence is about. The fronted item can be the grammatical subject, object or an indirect object (marked by the dative case). Thus, for example, the sentence in (115a) is about John who introduced Imre to Susan; the sentence in (115b) is about Susan such that Imre was introduced to her by John, and finally, the sentence (115c) talks about Imre who was introduced to Susan by John. What this means is simply that in each of the sentences the fronted item, which is the *topic* of the sentence, has the discourse function of expressing the central point of the sentence which also anchors the flow of the discourse in which the sentence concerned is embedded. It is this discourse function of being the topic that determines the order in which word categories serving specific grammatical functions (being the subject or the object, etc.) appear. To put it in another way, it is not the grammatical functions that determine which words or word categories will come first and which ones later, as it is in English in which the subject comes first, followed by the verb and then by the object. Rather, this particular role is played by localized discourse functions that have taken over the responsibilities borne by grammatical functions in other languages.

What is noteworthy in this connection is that the regulation of word order by discourse functions rather than by strict grammatical functions tends to make the word order in the language under consideration relatively more flexible. This suggests that any amount of flexibility that is introduced into the system of word order in a given language always arises due to certain factors that have some deeper links to grammatical functions in the language. This point can also be explicated by taking some examples from Japanese, a language that also favors word order freedom to a large extent. Let's look at the following examples.

(116)
a. Taroo–ga piza–o tabeta
   Taro–NOM pizza–ACC ate
b. Piza–o Taroo–ga tabeta
   pizza–ACC Taro–NOM ate
   'Taro ate pizza.'   (Miyagawa 2003)

As can be observed, both (116a) and (116b) convey the same thing more or less, although the word order is different. Now if we dig deeper into the reasons for this variation in the word order for virtually the same semantic content, it turns out that there are discourse functions at play. While (116a) is about Taro who ate pizza, the sentence (116b) is about the pizza that Taro ate. The second sentence gets the object to serve the discourse function of being the topic. In fact, there is an independent argument that this case is parallel to the English example in (114) above since in both cases an argument of the verb concerned is shifted from its place fixed in another construction (see Miyagawa 2009). Now the fact that freedom in word order can be achieved by means of discourse-functional strategies, irrespective of whether or not such strategies are grammatically marked, deserves independent attention. This is because variations in word order need to be somehow tracked in order that language users can plainly understand who did what to whom in any event or state of affairs. Once this is done, much of the complexity inherent in word order can be accounted for in terms of the *constitutive* properties of the underlying cognitive structures.

It is not just Japanese that allows for a flexibility in word order, there are South Asian languages that do behave the way Japanese does, although these languages differ from Hungarian in not having a *grammatically* fixed position for the fronting of any of the arguments of the verb. The following examples are from Bengali, a South Asian language spoken in India and Bangladesh.

(117)
a. tara pratidin Roy-ke dekhte ja-i
   they every day Roy-ACC see-NONF go-3PERS
   'They (go and) visit Roy every day.'
b. Roy-ke tara pratidin dekhte ja-i
   Roy-ACC they every day see-NONF go-3PERS
   'They (go and) visit Roy every day.'
c. pratidin dekhte ja-i tara Roy-ke
   every day see-NONF go-3PERS they Roy-ACC
   'They (go and) visit Roy every day.'

What the data from Bengali show is that the subject, the object, and the verb can all be moved around without resort to any grammatically licensed fronting position. Rather, the subject, the object and the verb undergo permutations in order to exploit a discourse-functional effect without thereby resting on grammatically marked positions. That is to say that when the object is fronted in (117b), for instance, the fronting takes place not because the front position is uniquely fixed by grammar for the displacement of arguments of the verb, as it is in Hungarian, but because the item fronted gains *focus* simply by virtue of being fronted.[6] That this is so is evidenced by the fronting of the verb itself in (117c), which indicates that the material minus the item fronted (i.e., the expression 'tara Roy-ke') does not form a separate or independent predicational structure, as it is in Hungarian.[7] In many ways, the behavior of fronting in this language is orthogonal to that in Japanese where the fronting takes place in order to raise the item to the status of the topic the sentence is all about, whereas in Bengali the fronting brings the item into focus. In short, the Japanese example in (116) is an example of what is typically called *topicalization*, whereas the Bengali example in (117) is an example of a focusing strategy, although both strategies have clear and distinctive discourse-functional impacts. Similar possibilities are also manifest in many Australian languages. Warlpiri, an Australian aboriginal language shows the typical patterns.

(118)
  a. karnta-ngku ka   yarla karla-mi
     woman-ERG PRES yam dig-NONPAST
     'The/a woman is digging yams.'
  b. yarla ka   karla-mi   karnta-ngku
     yam PRES   dig-NONPAST woman-ERG
     'The/a woman is digging yams.'
  c. karla-mi   ka   karnta-ngku   yarla
     dig-NONPAST PRES woman-ERG yam
     'The/a woman is digging yams.'   (Hale 1992: 64)
     [NONPAST = a nonpast tense marker]

It is quite evident here that Warlpiri exhibits word order variation in ways that are akin to those found in South Asian languages. Besides, other permutations in word order than the ones shown in (118) are also

---

[6]The underlying assumption here is that the basic word order in Bengali is SOV (Subject-Object-Verb), and hence, if the object appears at the beginning of the sentence rather than before the verb, the assumption is that the object has been fronted. Likewise, we say the verb has been fronted and so on. Note that if there is no basic word order present in a language, the concept of fronting does not even make sense (see Mithun 1992).

[7]That is the reason why in the Hungarian examples (115a–c) the material minus the item fronted is within braces.

possible in Warlpiri, which indicates that the language at hand exploits various discourse-functional strategies to such an extent that the concept of a *basic word order* in Warlpiri breaks down. This is another way of saying that word categories undergo permutations not because there are grammatically licensed positions reserved for particular word categories going through permutations but because different possibilities in word order variation reflect different demands or tendencies of discourse functions (such as what is newsworthy or more prominent and what is not).

There are, of course, various other linguistic idiosyncrasies and complexities in the word order of many languages that have been studied so far. What is significant for our primary concern is that much of the linguistic variation in word order is crucially structured around two modes of organization of words and/or word categories reflecting their cognitive structuring. The first type of ordering is the strategy whereby words and/or word categories are sequentially positioned along a linear dimension on which some item comes before some other item. In this ordering strategy, the sequential position of each item matters a lot for the assignment of not just grammatical functions but for discourse functions as well. Thus, for instance, if A precedes B and B precedes C (when A, B, and C are word categories), A's identity and its individuation depends principally on its being located before B and C. Similarly, B's individuation hinges on the fact that it follows A but precedes C, and C's individuation rests on the fact that it follows both A and B. In such a system of ordering, it is the sequential position that determines its relation to other items in the sequence and also the functional role assigned to it by virtue of its sequential position itself. Languages like English, French or Chinese fall under this category as difference sequences convey different meanings precisely because a difference in syntactic positions triggers a concomitant difference in the meaning. That is the reason why in English the sentences 'Rob loves his cat' and 'His cat loves Rob' mean two different things. Let's call this strategy, in a broader sense, *sequencing*. Sequencing is a kind of strategy or mode of cognitive organization that helps position items in specific locations primarily on the basis of their positions in a given sequence. Let's get the points straight here. Sequencing is not conceived of as a mere linguistic strategy; rather, the idea postulated is that whatever is manifest or instantiated as a fixed word order in languages like English or French reflects, or is a product of, a specific part of the underlying cognitive organization. The particular cognitive structuring responsible for what is manifest or instantiated as a fixed word order is sequencing.

Now distinct from sequencing is another type of cognitive structuring responsible for deviations from a fixed or rigid word order system. This strategy consists in rearranging items in different possible permutations and tracking the items through their markers that help individuate them in various permutations. Such a strategy does not simply rearrange words or word categories by means of permutations—this particular strategy is a kind of tracking strategy as well as it tracks the items that are reordered through permutation possibilities. Let's call this strategy *indexing*. Indexing is distinct from sequencing in the sense that sequencing rests on the positional information of a range of items in a sequence, while indexing does not have to deal with positional information as it tracks the items by tracing the specific individual markers that tag the items in question. Thus, indexing as a type of cognitive organization of a range of items the number of which goes beyond a reasonably demarcated bound[8] fixed by the mind helps operate on the items by utilizing the indices of the items concerned. These indices could be of various kinds. In the case of word order, the relevant indices are nothing but case markers or markers that indicate the roles of the particular arguments of a predicate, person/gender/number markers and also tense/aspect/mood markers placed on verbs. Languages like Japanese, Hindi, Bengali, Warlpiri, Hungarian, Russian, Spanish,

---

[8] It may be noted that the notion of a bound assumes a different level of significance, especially when the number of items is small enough. If this number is just two, for instance, the difference between sequencing and indexing is virtually inessential as they can be tracked either by their sequential positions or by their indices in more or less similar ways. Here, it does not make much of a difference since the permutation possibilities and the number of items are the same (that is, 2), regardless of whether sequencing or indexing is deployed. A greater number of items creates a great deal of burden on the memory and indexing can facilitate tracking many items within the space of a greater number of permutation possibilities which increase with the increasing number of items to be tracked. But sequencing, on the other hand, can be cumbersome for the memory if more than one sequence may be valid or licit for a given range of items whose number is greater than two.

Finnish, and many other African and Asian languages fit into this type. These languages have relatively free word order systems mainly because the notion of order in such languages does not make much sense. This is so because words or word categories are individuated and tracked by examining their linguistic indices which can be case markers, person/gender/number markers and tense/aspect/mood markers. These indices help track words or word categories, regardless of the order in which they appear, thereby making sequential information redundant. If sequential information is thus unnecessary in indexing, the very concept of order assumes a completely different form. Either it is instantiated through different permutation possibilities which need not themselves be tracked at all, or it is re-stated through certain reference positions, that is, positions that serve to act as reference points for certain word orders but not others. This requires more elaboration which is what we turn to now.

First, when indexing is realized in natural languages, free movement of words or word categories is easily accounted for. But it is not always the case that word order becomes absolutely free just because indexing is instantiated in natural language. That is to say that from the fact that indexing is instantiated natural language it does not follow that all possible permutations in natural language will also be implemented. Thus, the instantiation of indexing is, in fact, compatible with many possible scenarios which are indicative of the variation across free word order languages. We have observed that there is some amount of variation within and across free word order languages. There are languages like Warlpiri which permit free permutations, while there are others such as Hungarian that allow for word order flexibility within certain domains. Further, there are languages like Japanese, Hindi, etc. that allow for flexibility but this flexibility spreads out from a certain basic configuration of word order. With indexing in place, this variation can be accounted for once we recognize that indexing can not only give rise to free permutations in word order but also generate certain reference positions with respect to which permutations can be executed. A typical example of the latter possibility is clearly seen in Hungarian which fixes the front position as a reference position where words or word categories are freely shifted, thereby leading to flexibility in the word order but with respect to the front position as a general case. Warlpiri, on the other hand, constitutes another extreme case by not creating any reference position, and hence, free permutations occur without there being any local reference point.

Second, when reference positions are created within free word order languages, these reference points need not be just sequential positions. A referential point can be an entire sequence itself. A good example of this comes from those languages that have a basic word order configuration. For example, languages like Japanese, Hindi, etc. have SOV (Subject-Object-Verb) as the basic word order configuration. In such languages, this sequence constitutes the *unmarked* sequence—a sequence which is the most basic word order configuration against which other deviations or variations in word order are characterized. Once indexing is in place, an entire sequence can be marked as a reference point with respect to which sequence permutations within and across clauses are performed. Free word order languages that have a basic word order configuration can fix a given sequence as a reference point, and then, word order variations are defined with respect to that word order sequence which constitutes a reference point for deviations from the basic word order. This leads such languages to freely move words around within and across clauses. An example of this comes from the following Hindi sentences in which there are two finite (tense-marked) clauses with one being embedded within another. The relevant permutation takes place *from within* the embedded clause in (119).

(119)
a. Raju chah-ta hai ki tum Rinu-ko parao
   Raju want-3PERS-SING PRES COMPL you Rinu-ACC teach
   'Raju urges that you (should) teach Rinu.'
b. Rinu-ko Raju chah-ta hai ki tum parao
   Rinu-ACC Raju want-3PERS-SING PRES COMPL you teach
   'Raju urges that you (should) teach Rinu.'
                                    [COMPL = complementizer]

One important point to be taken into account in this connection is that even though languages such as Hindi, Japanese allow such permutations from within finite clauses, that is, across clauses boundaries, not all free word order languages are alike in this respect. Free word order languages that do not in the first place create any reference point in terms of a word sequence *may* or *may not* allow such across-the-clause permutations. Warlpiri and Mohawk independently give rise to the contrast in question even though both are free word order languages and do not create any reference point in terms of a word sequence. On the one hand, Warlpiri does not allow such across-the-clause permutations to take place,[9] Mohawk, an Iroquoian language spoken in parts of North America, on the other hand, does permit such across-the-clause permutations even if it has no basic word order (see also Sabel and Saito 2005). It is clear that creating a reference point in terms of a word sequence is concomitant with the implementation of across-the-clause permutations, but not creating in the first place a reference point is concordant with two possibilities—either the accommodation of across-the-clause permutations (as in Mohawk) or the rejection of such permutations (as in Warlpiri). Why is this so? To understand why this happens we have to dig deeper into the nature of indexing itself.

When indexing obtains, a certain reference point can be created so that word order variations may be instantiated with respect to that particular reference point. We have now understood that this reference point can be a sequential position or a whole sequence. A whole sequence, conceived of as a reference for sequential permutations, works well across clause boundaries in some languages (such as Hindi) because the sequential information extracted from the reference sequence adds in the information about the linguistic index of the item displaced, for example, the ACC case marker in (119). This more than facilitates the

---

[9]This particular aspect becomes more prominent in question formation requiring the permutation of the word that is questioned. The example below from Warlpiri is illustrative here.

nyarrpa Japanangka wangka-ja [pirrarni-rli   kuja   nyiya luwa-rnu]?
how Japanangka say-PAST [yesterday-ERG COMPL   what shoot-PAST
'What did Japanangka say he shot yesterday?'   (Hale 1994: 204)

Here the object of the verb 'luwa' (meaning 'shoot') is questioned but it does not undergo a permutation for question formation; instead, a proxy question word—that is, the word 'nyarrpa'—is placed at the beginning of the sentence to form the appropriate question.

shifting of words or word categories across clause boundaries. But when indexing is not accompanied by, or does not in itself lead to, the emergence of a reference point in the first place (i.e., a basic word order), a free word order language has to enrich the system of indexing further for certain words or word categories (arguments of predicates, for example) in a quite different way so as to permit them to go freely in and out of clauses. This is so because an enriched system of indexing allows for greater simplicity in the recognition of sentence constituents. A suitable example from Mohawk shows this quite clearly.

(120) atyatawi   ra-nuhwe's Sak
      dress   he/it -likes Jim
      'Jim likes the dress.'   (Baker 2001: 102–103)

It is evident from (120) that the verb itself carries specific markers bearing information about the subject and the object. Note that the arguments of the verb—the nouns 'atyatawi' and 'Sak'—do not bear any markers. This simply means that they can be shifted almost anywhere or even be dropped because the verb carries the necessary information. But when the relevant nouns carry case markers (as in the Warlpiri example (118), for instance, where the subject carries the ergative case marker), they cannot (always) be dropped since this will erase the necessary indices. These nouns may not undergo long-distance permutations in such a case, precisely because the verb does not *already* carry or encode the relevant information about the arguments. Consequently, this could lead to a confusion or ambivalence in the interpretation of the events expressed in multiple clauses as an argument undergoing across-the-clause permutations may end up being linked to the verb of the main clause rather than to the verb of the embedded clause. This is because the individual verbs in different clauses have not kept a record of the linguistic indices associated with their own arguments. Overall, this indicates that an absence of reference points (in terms of sequential positions or whole sequences) created in the deployment of indexing *can* be balanced by the exploitation of indexing in ways that preserve or 'store' the linguistic indices of certain items that can partake of across-the-clause permutations. But this must not be a universal choice, as the case of Warlpiri demonstrates that it can simply generate linguistic indices without leaving behind a record of such indices in the verb itself. That suggests that Mohawk is a conspicuous case of another *type of reference point* which is in the form of an encoding of indices that can be located within verbs. It is with *reference* to such encoding of indices localized in the verb that the items carrying such indices can freely go beyond clause boundaries.

With this understanding of the substantive differences between two types of cognitive structuring of a range of items as structurally expressed in natural language, we can now lay out the differences between sequencing and indexing in more precise terms.

(121) $<A_1, \ldots, A_j> \dashrightarrow R_1, \ldots, R_n$ Sequencing
(122) $S = \{I_i, \ldots, I_k\} \dashrightarrow R'_1, \ldots, R'_m$ Indexing
[Here it is not necessary that $i=j=n=k=m$, and '-->' means 'leads to']

The formulation in (121) states that there is a range of items $A_1, \ldots, A_j$ which form an *n*-tuple by virtue of the fact that they are in a sequence, and this leads to the construction of a number of relations $R_1, \ldots, R_n$ that can be defined over the items $A_1, \ldots, A_j$. That is to say that the items $A_1, \ldots, A_j$ when remaining in a sequence give rise to certain relations that define their formal relationships among themselves. In fact, it is easy to define precedence/succession relations over $A_1, \ldots, A_j$ and thus faithfully describe their sequential form. On the other hand, (122) defines a set S which consists of indices $I_i, \ldots, I_k$ that may be marked over $A_1, \ldots, A_j$, and then states that this set leads to the construction of a number of relations $R'_1, \ldots, R'_m$ (different from $R_1, \ldots, R_n$) that can hold among the indices themselves. It is these new relations $R'_1, \ldots, R'_m$ that the indices enter into which determine how a *reference point* in terms of a sequential position or a sequence or even an encoding of indices will be created. Thus, for example, when a sequential position is fixed as a reference point for certain word categories bearing the relevant indices, only those word categories having the relevant indices that are related to each other in an appropriate way (subject and object nouns, for example) can shift to the reference point. Topicalization is such a case in point. The indices $I_i, \ldots, I_k$ can enter into certain relations among themselves when they attach to the relevant words or word categories that are individuated and so distinguished by these indices. For instance, the accusative case marker differs from the ergative or nominative case marker, and hence, they represent different indices. Even the presence of a case marker on a certain nominal expression and its absence on another nominal expression lead to the creation of two different indices that entertain a relation of difference. What the formulations in (121–122) make clear is that sequencing as a cognitive structure has to start with the construction of an *ordered* formal structure while indexing has to construct an *unordered* system of indices which can be distributed over a range of items.

Sequencing and indexing are, therefore, two significant modes of cognitive structuring that underlie the system of word order across and within languages. The patterns of word order systems are not thus arbitrary regularities that more often than not tend to be found within and across languages. Word order systems are *in themselves* specific patterns of cognitive structuring which are disclosed once we go beneath the linguistic complexities and idiosyncrasies of word order. In fact, as the discussion above has attempted to show all throughout, the linguistic complexities and idiosyncrasies of word order also carry the marks of their cognitive structuring. The regularities unveiled easily fall out of the basic differences between sequencing and indexing, whereas other peculiarities and idiosyncrasies originate from the ways in which sequencing and indexing can be individually manifested or expressed in natural languages. But this is not to say that all the facts about word order in natural languages can be accounted for by appealing to sequencing and indexing any more than all facts about events involving gravity can be accounted for by appealing to physical laws and principles of gravity. The modest goal here is to simply show that word order systems can be properly conceived of as patterns of cognitive structuring when linguistic structures are themselves taken to be cognitive structures. This further advances the case for the cognitive structuring lying beneath linguistic structures.

4.1.5 Two Types of Grammar

The cognitive structuring underlying linguistic structures can be further revealed by examining not linguistic structures per se but distinct *representations* of linguistic structures. Linguistic structures are such that they cannot be observed in their entirety without the aid of certain notational representations. This is more so because sounds, meaning, and syntax are all organized into a seamless whole in a linguistic structure. While sounds are concrete entities, syntax and semantics are relatively more abstract by virtue of that fact that they have to be assimilated from the underlying organization of linguistic structures. This makes for a representation of linguistic structures in certain notational formats so that the structural organization can appear quite perspicuously. Once linguistic structures are couched in certain representational formats, the relevant distinctions and properties of linguistic structures that are structurally expressed become vivid and evident. This enables a smooth transition to the unmasking of aspects of cognitive structuring associated with the representations of linguistic structures. That is to simply say that certain aspects or dimensions of cognitive structuring linked to linguistic structures come into view only when the representations of linguistic structures are themselves examined minutely. This is exactly what we shall look into below.

Two types of grammars exist in current linguistic theory, and they are, by and large, grounded in the distinction of formats in which linguistic structures are represented. One is the derivational type of grammar, and the other is the representational type. Here, types of grammar are taken to be different representational systems in which linguistic structures are represented, analyzed, and so described. Grammars are thus ways of systematizing and uncovering aspects of linguistic structures which can be illuminated when grammar types impose an organization of their own on a variety of linguistic structures. The derivational type of grammars is based on the *structural* distinction of linguistic structures, whereas the representational type of grammars is grounded in the *ontological* distinction of linguistic structures (see also Mondal 2014). The ontological distinction of linguistic representations is motivated by (unique) differences in alphabets and specific principles, while the structural distinction of linguistic structures is anchored in differences in the nature of formal operations that apply to linguistic structures and can have empirical consequences. In this sense, the structural distinction of linguistic representations may, of course, feed the ontological distinction of linguistic representations. For example, in the Minimalist model of Generative Grammar (Chomsky 1995) syntactic operations output representations that are mapped onto logical form and phonological form as the former contributes to interpretations of meaning and latter to the articulation or perception of linguistic structures. The structural distinction of linguistic representations has been significant to the extent that structural transformations alter the representation of structure at one level giving rise to another level that contains the altered representation of the structural output. On the other hand, the ontological distinction of linguistic representations has been of considerable substance, in that some unique principles with unique alphabets operate in ontologically different domains or modules of language. Thus, phonology has phonemes, syllables, tone, etc. and principles like *Maximum Onset Principle* (that prefers onsets to coda in syllables), etc., while syntax has words, phrases, clauses, and a number of syntactic principles (for instance, *Extended Projection Principle* for the obligatoriness of subjects).

The difference between the ontological distinction of linguistic representations and the structural distinction of linguistic representations is substantial, insofar as it is responsible for splits in types of grammars. For instance, Lexical Functional Grammar (Bresnan 2001) has taken the ontological distinction of linguistic representations seriously because it has split linguistic form into C(constituent)-Structure and F(functional)-Structure, the former being syntactic and the latter being semantically grounded. Jackendoff's Parallel Architecture (2002) has also pursued the same path in keeping syntax different from both semantics and phonology, and vice versa, although there are connecting interfaces among them. A similar tendency is reflected in Dependency Grammar (Tesniére 1959) in which the dependency links that are semantic in nature form a wholly independent organization of linguistic structure which can be mapped onto the constituent structures of syntax. Head-Driven Phrase Structure Grammar (Pollard and Sag 1994) and Cognitive Grammar (Langacker 1987, 1999) have, on the other hand, conflated both distinctions into a single unified level or system of representation. Although the structural distinction of linguistic representations is considered fundamental in Generative Grammar, it would perhaps be wrong to say that it has taken the ontological distinction of linguistic representations blithely. In fact, Generative Grammar has always oscillated between these two relevant distinctions (Anderson 2006). A simple example can illustrate the basic differences between the derivational type of grammatical encoding and the representational type of grammatical encoding.[10] This is shown in Fig. 4.11.

Figure 4.11 shows that the sentence 'The girls flirted with the boys' can be represented in two different ways in the two grammatical types. Within the derivational type of grammatical representation, the structure of the sentence 'The girls flirted with the boys' as we see it is *derived* from another level of structure where the subject 'the girls' is placed within the *v*P and the main verb remains in its original position. Certain displacement operations alter this level of structure, yielding a structure in which the subject occupies the topmost position within the TP because it requires case and the main verb attaches to the light verb because certain verbal features of the light verb match those of the main verb. These displacement operations are not always transparent to the surface structure of sentences as we see it, and that's why the sentence string 'The girls flirted with the boys' matches not just the structural representation at the pre-derivational stage (when structural alterations have not taken place) but also the structural representation at the post-derivational stage (the structure after the structural alterations have taken place).[11] The representational type, on the other hand, shows only

---

[10]The glimmering of this difference was perhaps recognized in the formulation of Generative Semantics (Lakoff 1971; Postal 1972) wherein the semantic representation from which syntactic surface structures were derived was representational, whereas the emergence of surface structures was derivational.

[11]One scenario where displacement operations can be transparent to the surface structure is the case of raising. The following shows this clearly.

    Sentence (S): The professor seems to be smart.
    Pre-Derivational Representation: [$_S$ seems [[the professor] to be smart]]
    Post-Derivational Representation: [$_S$ [the professor] seems [__ to be smart]]

4 Linguistic Structures as Cognitive Structures 219

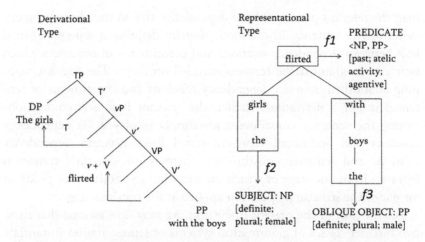

Fig. 4.11 The derivational and representational types of grammatical representation (TP = Tense Phrase, vP = Light Verb Phrase, VP = Verb Phrase, DP = Determiner Phrase, PP = Prepositional Phrase. The arrows in the tree diagram under the derivational type of grammars represent displacement operations, and *f1* ... *f3* under the representational type represent different functions that map elements of dependency tree to phrasal constituents)

the semantic links among the constituents of the sentence in question. Figure 4.11 shows that the head of the subject noun phrase, that is, 'girls' depends for its status as the head of the subject phrase on the verb 'flirted,' and so does the preposition 'with' for its status as the head of the prepositional phrase located within the verb phrase. Likewise, the determiner 'the' depends on the head 'girls' for its status as the determiner that specifies the meaning of the noun head 'girls,' and the noun head 'boys' depends on the preposition 'with' for its status as the argument of the preposition 'with.' The functions f_1, f_2, and f_3 in Fig. 4.11

The symbol '_' above indicates the place from where the noun phrase 'the professor' has been shifted. This structural displacement has taken place because the noun phrase 'the professor' does not get case within the infinitival clause 'to be smart' (which is tenseless) and hence moves to the front in order to receive case. The point to be highlighted is that the string 'The professor seems to be smart' matches the structural representation at the post-derivational stage, while it differs from the structure at the pre-derivational stage.

map the relevant portions of the dependency tree to the phrasal constituents of the sentence in question, thereby defining a representational link between dependency relations and constituents of sentences which bear certain grammatical features encoded on them. The relevant mapping established between dependency relations and constituents of sentences is not a derivation. Rather, the relevant linguistic connections among the sentence constituents are simply displayed via dependency relations and additional representational links between dependency relations and sentence constituents. There is no structural transition involved from one stage of structural alteration to another that yields an output of the structural operation applied at the previous stage.

For the sake of the present discussion, we may now assume that these two broader classes of grammatical systems of representation instantiate two conceptually distinct types of grammars. Given that the two types of grammars are oriented around distinct principles, it appears that they must have different formal consequences for the relation between language and cognition. Closer inspection reveals that these two grammar types may prove equivalent to one another in certain linguistic contexts, but not others. Figure 4.11 shows that the derivational type of grammars and the representational type of grammars are equivalent, inasmuch as both types articulate the structures of certain constructions in essentially similar ways, as evidenced by the diagrammatic representation of the sentence 'The girls flirted with the boys' in the figure. What is, however, more interesting is that these two types of grammars diverge, especially when structure-changing operations relate a number of constructions to each other on the one hand (the derivational type preferred), and *intra*-constructional elements instantiate certain conceptual or structural relations with respect to one another on the other (the representational type favored). The following examples show this quite succinctly.

(123)
a. It is hard to understand why he did it.
b. Why he did it is hard to understand.
(124) These animals are our best companions.
(125) Max and Sax love each other.

While (123) represents the case where structure-changing operations underlie the linguistic relation between (123a) and (123b),[12] (124–125) are cases in which a structural relation between two or more items (as in (124) where the plural agreement between 'these,' 'animals,' 'are,' and 'companions' is involved) or a referential relation (as in (125) where 'Max and Sax' and 'each other' are co-referential) is captured better in the representational type of grammars which employs mappings between feature sets, trees, and signs.

In view of these considerations, it is compellingly clear that the two types of grammatical representation are formally equivalent in certain linguistic contexts, but spell out distinct consequences in other contexts involving relational linking, reference, and interpretation of signs (as in (123–125)). This implies that the two types of grammatical representation thus have distinct levels of cognitive organization, and hence, there exists a cognitively grounded differentiation between them, leading to divergent consequences which possibly reflect two different ways in which the mental organization of grammar projects itself onto the world. That is, the cognitive differentiation between the two types of grammatical representation provides evidence for distinct mental modes of (re)presentation of grammar with respect to their symbolic significance that reflect two different aspects or *dimensions* of the mental organization of grammar. This uncovers one significant aspect of cognitive structuring with respect to linguistic signs. Linguistic structures are constituted by linguistic signs (words, phrases, clauses, etc.), and these structures can be represented in distinct encoding systems of grammatical representation. It turns out that distinct encoding systems of grammatical representation tap distinct properties or dimensions of linguistic structures and thus turn on distinct structuring strategies of the cognitive system. Let's ponder over it for a moment.

---

[12]Here the clause 'why he did it' can be said to originate in the argument position of 'understand', as it is in (123a), and since the subject position of the entire sentence is empty, a dummy subject (which is 'it') is inserted in the front position. But then, the entire clause 'why he did it' can itself play the role of the subject of the whole sentence and may then move to the front position to serve as the subject. This is what we see in (123b).

First, when two types of grammatical representation are (more or less) equivalent to each other in their expressive power by virtue of the fact that they express the relevant structural properties of linguistic structures in essentially the same way (as in Fig. 4.11, for example), neither of the two types of grammatical representation can be considered superior to the other in the representation of the relevant linguistic structures. But, if and when one of the two types of grammatical representation turns out to be superior to the other in expressing and representing certain linguistic properties and/or facts, it is clear that the two types of grammatical representation are not actually equivalent to one another as far as those particular linguistic structures are concerned. Now the postulation here is that the equivalence of the two types of grammatical representation obtains due to the way the cognitive system *can* organize linguistic structures with the aid of its representational or encoding resources. When two types of grammatical representation encode the relevant linguistic properties in essentially the same way, the underlying cognitive structuring displays a kind of duality in being open to two modes of representation—the derivational type and the representational type. It may be observed that the derivational type consists in the specification of a number of transitions from one stage of structural representation to another. Simply speaking, the derivational type is a *dynamic* or procedural system of representation for linguistic structures, whereas the representational type is a static or declarative system of representation for linguistic structures. If this is the case, the cognitive system can allow for both dynamic and static representations for the same class of linguistic structures. That is to say that there exists some form of *representation invariance* for (certain) linguistic structures at a certain level of the mind, and this accounts for the emergence of the duality of representational formats permitted by the cognitive system.

An analogy can clarify the matter at hand in a more illuminating way. Consider chess, for instance, to be a symbolic system of rules. On the one hand, its rules can be completely described or encoded by means of a set of declarative statements. But, on the other hand, its rules can also be specified by drawing up all possible moves of chess pieces in terms of diagrams or something of that sort. Clearly, the former is a static representational format, whereas the latter is a dynamic representational format. Both representational formats are expressively equivalent to each other, and hence, either of them can serve to characterize the system of rules of chess. But where does this equivalence come from? Does it solely emerge from the symbolic system itself? Since no symbolic system can on its own invent or give rise to its own interpretations, it must be the case that the required interpretation has to be channeled along certain lines or modes of representation made available by the cognitive system at hand. Unless the mind is capable of organizing the same symbolic system in two modes of representation, it is not clear how this could ever be possible. Therefore, the cognitive system itself must have it its repertoire two ways of encoding or representing the same things. Thus, the duality of representational formats emerging from a symbolic system has to be attributed to the cognitive structuring capacities inherent in the mind. In the present case, distinct notational or symbolic representations of linguistic structures do tell us something non-trivial about the representational capacities of the mind just as distinct modes of representational formats that can be evoked for the encoding of chess rules tell us something about the mind's ways of framing a set of symbolic rules.

Second, we have so far concentrated on the duality of ways in which the mind can organize and encode structural properties of linguistic structures. But we have also observed that there are certain cases in which the duality breaks down and the one particular representational type proves to be expressively superior to the other type in expressing structural properties of linguistic structures. This case is the other side of the same coin. More precisely, when two types of grammatical representation diverge in having distinct consequences for (certain) linguistic structures, each type assumes a cognitive significance of its own by differentiating itself from the other type. The collapse of the duality of the two types of representation occurs due to the emphasizing of the features and aspects of each type in the encoding of certain linguistic structures. To put it in other words, the structural properties of certain linguistic structures themselves help emphasize the features and aspects of either of the two types of representation schemas. This thereby ends up magnifying the cognitive significance of one type of representation schema rather than the other one in a given scenario. But where does this cognitive differentiation originate from in the first place? Surely this simply cannot be attributed to the linguistic structures in themselves, as also pointed out above. A type of representation schema becomes consolidated only when its properties find their base in those symbolic structures that bear those properties. Thus, (123) is an example where the derivation type of representational schema is consolidated and thus distinguishes itself from the representational type, simply because the linguistic structures in (123) bear the exact properties that characterize or constitute the derivational type. Likewise, examples (124–125) bear the exact linguistic properties that are inherent in the representational type. This indicates that types of representational formats can diverge when they have distinct consequences for the cognitive system. This requires elaboration. It may be noted that the derivational type and the representational type are *formally* equivalent to each other when applied to those linguistic structures that are neutral with respect to the *unique* features of either of the representational schemas. But there are other linguistic constructions that are more compatible with either of the two types of grammatical representation. If this is so, it is evident that one of the types becomes *cognitively loaded* in certain cases of linguistic structures but not in others in the sense that one type of representation schema rather than the other activates or evokes all the unique cognitive properties and features associated with that type. Thus, for example, when the derivational type in (123) turns out to be superior in encoding the structural properties of the linguistic structures, the dynamicity inherent in the transitions from one structural level of representation to another becomes cognitively more salient. Similarly, when the representational type in (124–125) proves to be more adequate, various representational devices (such as mapping functions or relations and feature matching) are immediately mentally instantiated and thereby become more cognitively prominent.

In fact, this particular tendency of representational formats to generate divergent consequences in certain contexts despite the fact that they are otherwise formally equivalent may ride on the *intensional* orientation of symbolic systems. The property of intensionality is best captured in contexts of belief when two terms have distinct cognitive consequences even though the terms refer to the same entity. An example of this can be provided here. Let's take the sentence 'The inventor of telephone must be a genius.' Now we know that 'Alexander Graham Bell' and 'the inventor of telephone' refer to one and the same person. Even though this is the case, it is quite possible that someone who is not aware of this fact may form two distinct beliefs about the terms 'Alexander Graham Bell' and 'the inventor of telephone.' Therefore, the substitution of 'the inventor of telephone' for 'Alexander Graham Bell' in the sentence 'The inventor of telephone must be a genius' may not hold true for the person who holds distinct beliefs about the two terms. Here is a scenario in which two terms are referentially equivalent but have cognitively distinct consequences due to the associated beliefs attaching to a given term. The point raised here is that something similar is going on in the case of types of representational formats being used to encode structural properties of linguistic structures. There is a kind of intensional incongruity between the derivational type and the representational type when they are used for the representation of linguistic structures. On the one hand, these two types are formally equivalent in being representationally adequate in more or less equal terms, but on the other hand they have distinct cognitive consequences when their unique cognitive properties become hooked to the aspects of linguistic structures consistent with those cognitive properties. This underpins the intensional character of the 'oscillation' between the two types of representational formats the cognitive system manipulates or tends to deploy. In other words, a cognitive incommensurability holds between the two types of representational formats and thus constitutes the fulcrum of intensional gap between them.

This has consequences for the distinction between an abstract system of grammar and the parsing grammar used in real time of language processing, which may also be aligned with the cognitive differentiation between the derivational type and the representational type. A distinction is often made between an abstract system of grammar internalized in the mind and a parsing grammar used for language processing in real time. This distinction appeals to differences between the form of the abstract system of grammar and the character of the parsing grammar which is a real-time system for actual language use. While the abstract system of grammar can allow for well-formed strings which are arbitrarily long, say, a million-word-long sentence, the parsing grammar cannot handle such strings because it relies on real-time cognitive resources such as memory and mental tokens of representations. This distinction is akin to the distinction between a 'competence grammar' and a 'performance' grammar (Chomsky 1965, 1985). However, this need not be taken to mean that there is a strict (ontological or substantive) separation between the two systems since they may often coincide with each other. Sometimes performance/parsing principles are integrated into the system of competence grammar in ways that indicate that the abstract system of grammar need not be entirely disjoint from the parsing grammar (see also Hawkins 2004). Thus, for example, pieces of linguistic structure that are not immediately adjacent to those linguistic elements to which they are syntactically/semantically related are often hard to process.

(126)
a. The minister brushed off the allegation that he bribed the foreign officials.
b. The minister brushed the allegation that he bribed the foreign officials off.

(127)
a. This writer in his novel has depicted the picturesque scenes of the countryside in a way that is not only splendid but also tantalizingly refreshing.
b. This writer in his novel has depicted in a way that is not only splendid but also tantalizingly refreshing the picturesque scenes of the countryside.

The difference between (126a) and (126b) is akin to that between (127a) and (127b), in that in both cases this shows that a piece of linguistic structure that is not immediately adjacent to what it is related to is hard to parse. Hence the sentences in (126a) and (127a) are taken to be more natural in English and thus integrated into the competence grammar. Similar cases have also been discussed in examples (11–14) in Chapter 3. While parsing principles can be integrated into the abstract system of grammar, parsing principles, although guided by the abstract

4 Linguistic Structures as Cognitive Structures

internalized grammar, can act independently of the abstract system of grammar (Pritchett 1992). The following sentences belonging to the *garden-path*[13] type illustrate the point well.

(128) Without her donations to the charity failed to appear.
(129) After Todd drank the water proved to be poisoned.
(130) Susan convinced her friends were unreliable.

All these sentences in (128–130) induce a difficulty in processing. In (128) this difficulty arises at the position of the word 'failed,' in (129) at the position of the word 'proved,' and in (130) at the position of 'were.' The difficulty may disappear once the relevant piece of linguistic structure is reanalyzed. Thus, when the noun phrase 'donations to the charity' in (128) is reanalyzed as the argument of the verb complex 'failed to appear,' the sentence begins to make sense. Similarly, if the noun phrase 'the water' in (129) is not analyzed as the object of 'drank,' the sentence turns out to be well-formed, and then in (130) the noun phrase 'friends' has to be interpreted as the subject of the embedded clause 'friends were unreliable.' This unequivocally suggests that the parsing system can behave in ways that are not fully predicted by the abstract system of grammar internalized in the mind. More importantly, this also demonstrates that the abstract system of grammar and the parsing grammar can be virtually indistinguishable from each other and at the same time may also diverge from each other in the ways that carry

---

[13]Some sentences are known to be grammatically well-formed and yet introduce processing difficulties. The following sentence first discussed by Bever (1970) is a paradigm example of this.

'The horse raced past the barn fell'.

This sentence introduces the difficulty when the reader/listener is sort of 'led down the garden path' in interpreting 'raced' as the matrix verb of the main subject 'the horse', when in fact the actual matrix verb is 'fell' which is related to 'the horse' but this relation is interrupted by the reduced relative clause 'raced past the barn'. That is, the sentence becomes clearer if we insert 'that was' between 'horse' and 'raced', thereby changing it to 'The horse that was raced past the barn fell'. Hence the name 'garden path' is attached to such constructions.

distinct consequences when they drift away from each other. The present proposal is that this disparity between the two systems of grammar may be due to an intensional gap between the two systems. This intensional gap is parallel to the one between the derivational type and the representational type of grammatical notations. When the two systems of grammar fold into each other and thus merge together, the underlying cognitive organization does not make any substantive distinction between the two. But, when they yield divergent consequences, they carry distinct cognitive consequences. On the one hand, the abstract system can permit infinite extensions by way of its abstraction from actual conditions of language use and does not appeal to processing factors (cognitive resource limitations, for example); the parsing grammar does operate by being governed by real-time limitations of the cognitive system at work and hence may even stumble over the outputs of the abstract system of grammar. This disjunction between the two systems may be rooted in an intensional gap at some level of our cognitive organization. This lines up well with the converging insights gained into the structure of cognitive organization looked through the constitution of linguistic structures.

4.2 Summary

This chapter has attempted to demonstrate that natural language in all its complexity has enough logical richness that helps reveal the internal cognitive substance linguistic structures are made of. A number of complex linguistic phenomena have been examined and analyzed with this goal in mind, yielding unique insights into the nature of their cognitive constitution. The overall picture that emerges is as enriching as it is indicative of the co-construction of language and cognition. Finally, this broadens the boundaries of the realm of possible cognitive structures as diversified patterns of linguistic structures open up a vast range of cognitive structures hidden beneath themselves.

References

Anderson, J. M. (2006). Structural analogy and universal grammar. *Lingua, 116,* 601–633.

Baker, M. C. (2001). *The Atoms of Language: The Mind's Hidden Rules of Grammar.* New York: Oxford University Press.

Baker, B., & Harvey, M. (2010). Complex predicate formation. In M. Amberber, B. Baker, & M. Harvey (Eds.), *Complex Predicates: Cross-linguistic Perspectives on Event Structure* (pp. 13–47). Cambridge: Cambridge University Press.

Baltin, M., Déchaine, R., & Wiltschko, M. (2015). The irreducible syntax of variable binding. *LingBuzz.* http://ling.auf.net/lingbuzz/002425.

Barwise, J., & Cooper, R. (1981). Generalized quantifiers and natural language. *Linguistics and Philosophy, 4,* 159–219.

Bever, T. G. (1970). The cognitive basis for linguistic structures. In J. R. Hayes (Ed.), *Cognition and the Development of Language* (pp. 279–362). New York: Wiley.

Bodomo, A. B. (1997). *Paths and Pathfinders: Exploring the Syntax and Semantics of Complex Verbal Predicates in Dagaare and other Languages* (PhD dissertation) Trondheim: The Norwegian University of Science and Technology.

Bresnan, J. (2001). *Lexical Functional Syntax.* Oxford: Blackwell.

Butt, M. (1995). *The Structure of Complex Predicates in Urdu.* Stanford, CA: CSLI Publications.

Chomsky, N. (1965). *Aspects of the Theory of Syntax.* Cambridge: MIT Press.

Chomsky, N. (1985). *Knowledge of Language: Its Nature, Origin, and Use.* New York: Praeger.

Chomsky, N. (1995). *The Minimalist Program.* Cambridge: MIT Press.

Fauconnier, G. (1994). *Mental Spaces.* Cambridge: Cambridge University Press.

Foley, W. A. (2010). Events and serial verb constructions. In M. Amberber, B. Baker, & M. Harvey (Eds.), *Complex Predicates: Cross-linguistic Perspectives on Event Structure* (pp. 79–109). Cambridge: Cambridge University Press.

Foley, W. A., & Olson, M. (1985). Clausehood and verb serialisation. In J. Nichols & A. C. Woodbury (Eds.), *Grammar Inside and Outside the Clause* (pp. 17–60). Cambridge: Cambridge University Press.

Fortuny, J. (2015). From the conservativity constraint to the witness set constraint. *LingBuzz*. http://ling.auf.net/lingbuzz/002227.
Gärdenfors, P. (2000). *Conceptual Spaces: The Geometry of Thought*. Cambridge: MIT Press.
Hale, K. L. (1992). Basic word order in two "free word order" languages. In D. L. Payne (Ed.), *Pragmatics of Word Order Flexibility* (pp. 63–82). Amsterdam: John Benjamins.
Hale, K. L. (1994). Core structures and adjunctions in Warlpiri syntax. In N. Corver & H. van Riemsdijk (Eds.), *Studies on Scrambling: Movement and Non-Movement Approaches to Free Word Order* (pp. 185–219). Berlin: Mouton de Gruyter.
Hawkins, J. A. (2004). *Efficiency and Complexity in Grammars*. New York: Oxford University Press.
Jackendoff, R. (1990). *Semantic Structures*. Cambridge: MIT Press.
Jackendoff, R. (2002). *Foundations of Language: Brain, Meaning, Grammar, Evolution*. New York: Oxford University Press.
Keenan, E. L., & Stavi, J. (1986). A semantic characterization of natural language determiners. *Linguistics and Philosophy, 9*(3), 253–326.
Kiss, K. É. (1998). Discourse-configurationality in the languages of Europe. In A. Siewierska (Ed.), *Constituent Order in the Languages of Europe* (pp. 681–728). Berlin: Mouton de Gruyter.
Lakoff, G. (1971). On generative semantics. In D. D. Steinberg & L. A. Jakobovits (Eds.), *Semantics: An Interdisciplinary Reader in Philosophy, Linguistics and Psychology* (pp. 232–296). Cambridge: Cambridge University Press.
Langacker, R. (1987). *Foundations of Cognitive Grammar*. Stanford: Stanford University Press.
Langacker, R. (1999). *Grammar and Conceptualization*. Berlin: Mouton de Gruyter.
Levin, B., & Hovav, R. (2005). *Argument Realization*. Cambridge: Cambridge University Press.
Mithun, M. (1992). Is basic word order universal? In D. L. Payne (Ed.), *Pragmatics of Word Order Flexibility* (pp. 15–62). Amsterdam: John Benjamins.
Miyagawa, S. (2003). A-movement scrambling and options without optionality. In S. Karimi (Ed.), *Word Order and Scrambling* (pp. 177–200). Oxford: Blackwell.
Miyagawa, S. (2009). *Why Agree? Why Move?: Unifying Agreement-Based and Discourse-Configurational Languages*. Cambridge, Mass.: MIT Press.

Mondal, P. (2014). *Language, Mind, and Computation*. London: Palgrave Macmillan.

Peters, S., & Westerståhl, D. (2006). *Quantifiers in Language and Logic*. Oxford: Clarendon Press.

Pollard, C., & Sag, I. (1994). *Head-Driven Phrase Structure Grammar*. Chicago: University of Chicago Press.

Postal, P. (1972). The best theory. In S. Peters (Ed.), *Goals of Linguistic Theory* (pp. 131–170). Englewood Cliffs, NJ: Prentice-Hall.

Pritchett, B. (1992). *Grammatical Competence and Parsing Performance*. Chicago: Chicago University Press.

Reuland, E. J. (2011). *Anaphora and Language Design*. Cambridge: MIT Press.

Sabel, J., & Saito, M. (Eds.). (2005). *The Free Word Order Phenomenon: Its Syntactic Sources and Diversity*. Berlin: Mouton de Gruyter.

Tesniére, L. (1959). *Eléments de Syntaxe Structurale*. Paris: Klincksieck.

van Benthem, J. (1983). Determiners and logic. *Linguistics and Philosophy*, 6(4), 447–478.

van Hoek, K. (1996). A cognitive grammar account of bound anaphora. In E. Casad (Ed.), *Cognitive Linguistics in the Redwoods: The Expansion of a New Paradigm in Linguistics* (pp. 753–792). Berlin: Mouton de Gruyter.

von Fintel, K. (1997). Bare plurals, bare conditionals, and only. *Journal of Semantics*, 14(1), 1–56.

Wilson, S. (1999). *Coverbs and Complex Predicates in Wagiman*. Stanford, CA: CSLI Publications.

5

Conclusion

This chapter will attempt to stitch together the threads of arguments running all throughout the book. Overall, the present book has attempted to demonstrate that the chasm between the scale/level of cognitive systems/structures and that of biological structures is too wide to be eliminated. There are a number of facets of the linguistic system that offer deeper insights into the structural format of cognition but cannot in any way be traced to the biological substance on the one hand, and biological structures (genetic structures and aspects of neurobiology) do not also attach to these facets of the linguistic system on the other hand. This has motivated us to look for the right locus of the structural organization of cognition. Significantly, a vast territory of the logical structures and patterns of cognition is uncovered through language without appealing to biological relations, which do not reasonably jell with the structural format of linguistic cognition in terms of either descriptive or explanatory adequacy. Therefore, a careful and judicious way forward is to accept and probe into the gaps in language-biology relations, irrespective of how one takes into consideration the matter of levels of organization for any given arena of inquiry. Nonetheless, the aspects of gaps need not be considered insignificant for an understanding of levels of intentional description in relation to levels of organization. Rather, any aspects of gaps that cannot be *logically* (rather than practically) reconciled with aspects of integration or even of unification may be located in different planes or dimensions of nature. In the present context, this means that the prospects of integration of the computational level with the algorithmic and implementational levels (in Marr's (1982) sense) for linguistic phenomena are indicative of merely the forms of epistemic and possibly methodological integration. The manifestations of the desirable integration in language-biology relations are merely epistemic and possibly methodological as well because the underlying strategies of unification consist in uniting the knowledge bases of the linguistic and neurobiological levels or scales and thereby showing possibilities of fruition of shared or common methods of investigations into both linguistic and neurobiological levels of organization of nature. This sort of strategies is reliably distinct from another sort of strategies that concentrate on finding out either a common cognitive architecture remaining invariant for many cognitive domains or a collection of elementary cognitive processes and/or structures that can give rise to complex forms of cognition across various domains of cognition (see Milkowski 2016). The former is notably vertical in virtue of moving up and down across levels or scales of description superimposed over each other for any single cognitive domain or module, while the latter is basically horizontal in involving unification of a number of different cognitive domains or modules (see for a related discussion, Roth and Cummins 2017). The consequence arising from this is that if any postulated facets of integration in language-biology relations are merely epistemic and methodological, the aspects of gaps revealed in this book must be ontological and in many respects epistemological. That is because aspects of gaps essentially reflect the special nature and form of linguistic structures on the one hand and the limits of knowability in the context of language-biology relations on the other. In other words, the nature of *being (neuro)biological* cannot automatically license inferences about the intrinsic nature of what is linguistically structured, and the nature of *being linguistic* is thus enclosed in terms of knowability off from the nature of being (neuro)biological. The concept of knowledge of language may thus not be the object of knowing when knowing is crucially based on inductive or abductive inferences from the grounding in (neuro)biology.

From a significant perspective, the division of labor between aspects of language-biology integration (say, in terms of actual linguistic functions such as naming and referring) and all aspects of gaps may be conceptualized in terms of Haugeland's (1978) notion of dimensions and levels. He attempted to account for cognitive activity in terms of dimensions that are superimposed over each other much like Marr's (1982) three levels but each of these dimensions can have a degree of variation in its level of intentional interpretation. Therefore, Haugeland's dimensions are vertical, whereas levels within a(ny) dimension are horizontal. The distinction can be illustrated with his example of chess programs. Actualized chess programs (in terms of their mode of functioning and usefulness) are situated on the highest dimension with the programs coded in a certain language (say, in LISP) being on the intermediate dimension and the logic circuit programs being on the lowest dimension. Now each of these dimensions can have varied levels of intentional interpretation depending on how systematic an explanation is of the valid chess moves regardless of the way the moves are executed (by means of lights, bar pressing or something else). This distinction between dimensions and planes of intentional interpretation on a given dimension is instrumental in the current context. The idea postulated here is that unity in language-biology relations (in terms of integration or otherwise) is to be achieved in terms of the unity of dimensions on which language is to be described or studied. These dimensions can be those of linguistic structures, psycholinguistic procedures and neurobiological mechanisms. But, on the other hand, any gap in language-biology relations must be located on varying levels of intentional interpretation on a given dimension. Thus, given any dimension a gap arises from the very fact that some properties or forms of language on that dimension cannot have or exhibit the level of intentional interpretation appropriate for the unity of dimensions. This happens because the ontological uniqueness of a certain facet of language allows one level of intentional interpretation but disallows another level required for the unity of dimensions. In the current context, we can thus state that the aspects of gaps explored in Chapters 1 and 2 are actually located at higher levels of intentional interpretation which cannot be targeted for the desired integration of dimensions for language-biology relations.

Hence these aspects of gaps thwart any postulated unity of dimensions from the higher levels of intentional interpretation located on each of the higher dimensions (i.e., on the dimensions of linguistic structures and psycholinguistic procedures). To put it in other words, the gaps remain located across all the dimensions covering all and only the higher levels of intentional interpretation, while the matters of integration may be able to target only the lower levels across all dimensions. This allows for the coexistence of biological constraints operating on lower levels of the manifestation of language (such as uttering, naming, and perception of linguistic sounds) and higher-level properties of language not revealed by biological investigations.

Therefore, even though biological relations are ultimately irrelevant to the logical character of linguistic cognition, the essence of the argument here does not, however, scale down the role of biological structures in the growth and evolution of cognition. Rather, its role of biological implementation, rather than of biology itself, in the present context has been *miniatured* but not certainly eliminated. The biological underpinning of language acquisition, the growth of grammar in adults, language loss and language disorders has certainly helped secure a firmer understanding of the role of biology in language. Without the biological scaffolding, the internalization of language cannot even get off the ground. Plus the development of linguistic cognition is often marked by trajectories that are due in large measure to the channeling of biological structures along 'canalized' paths. Likewise, various stages of language evolution have to be understood against the background of biological principles. This would doubtless show that biological constraints and principles guide, govern and shape processes of growth, development and instantiation of the language faculty as we understand it.

Nevertheless, this does not at the same time demonstrate that the internal symbolic elements of the cognitive system that language helps (re)structure are to be traced to, or modulated by, biological structures or properties pertaining to biological relations of instantiation. An insistence on sticking to the supposition that biological relations of instantiation can provide the right conceptual tool for understanding cognitive structures ends up advancing a much grander and specious argument.

This is an argument that is not fully supported either by empirical soundness or by conceptual clarity, as we have observed. The moderate position that can be maintained is that biological principles and constraints which also include biological relations of instantiation form the 'contours' of the cognitive architecture as it develops or evolves, but the internal contents or structures of the cognitive architecture cannot be characterized in the same way. The arguments marshaled in this book reinforce the idea that the internal structural contents of what we may characterize as linguistic cognition elude and in many ways defy any biological description that purports to have such contents traced or reduced to biological structures. Rather, they can be best explored from the vantage point of linguistic structures themselves. Significantly, the central line of reasoning employed in the book must not be taken to tally with the stance adopted by Carr (2009) when he asserts that the ontological realm of linguistic structures is independent of the realm of the cognitive and, for that matter, of the realm of what is biological. Carr maintains this position because he believes properties of linguistic structures are properties of a speaker-external or intersubjective domain, and hence they are irrelevant even to the cognitive domain. The present book does not espouse such an extreme *autonomist* position on the grounds that many otherwise (logically) abstract and thus speaker-external (because logical abstractions do not belong to any individual) properties of linguistic structures evince aspects of cognitive constitution, as is clearly shown in Chapter 4. As we have observed that many logical aspects of linguistic structures *are* aspects of underlying cognitive structures of which linguistic structures are themselves made up, there is reason to believe that there is more to linguistic structures than mere nebulous abstractions removed from the underlying level of cognitive organization. On the one hand, this adds to the richness of the cognitive system by showing that a description of the cognitive system simply in terms of its architectural constitution and the contours defining its boundaries barely scratches the surface of the cognitive substrate. We need to dig deeper into the cognitive system in order to uncover the formats of the structural contents lodged inside the cognitive system. On the other hand, language turns out to be the appropriate tool for achieving this goal as it defines and marks the realm of linguistic cognition and thereby offers the much-needed entry into the space of the cognitive machinery. This helps understand the internal constitution of linguistic structures that beefs up the account of complexities and idiosyncrasies of a host of linguistic phenomena from a fresh new perspective.

In a nutshell, the crucial message to be driven home is that it is easy to fall in with tempting biological descriptions of linguistic cognition, but one must be cautious about extending it way beyond its explanatory territory. Thus, we must recognize that biological accounts work best when they are applied within domains where the principles and constraints of biology are transparent to the character of the structures or forms studied. This is clearly evident for the study of wings, feathers, tails, internal organs of organisms, nervous structures, etc., to mention just a few. The prospect of an appropriate biological account of the cognitive structures assembled and assimilated in linguistic cognition is not simply impugned—it is also banished from the theoretical minimum required for an understanding of cognition in general. If this is what the texture of the cognitive reality looks like or appears to be, so be it. We do not need to press biological facts to suit any linguistic theory. Rather, no effort should be spared to penetrate into the realm of linguistic cognition for its own sake and for its sheer beauty of the inherent logical complexity. With this view in place, we may hope to make forays into the unexplored and shadowed territory of linguistic cognition in the quest for new vistas of language–cognition relations.

References

Carr, P. (2009). *Linguistic Realities: An Autonomist Metatheory for the Generative Enterprise*. Cambridge: Cambridge University Press.

Haugeland, J. (1978). The nature and plausibility of cognitivism. *Behavioral and Brain Sciences, 1*, 215–226.

Marr, D. (1982). *Vision: A Computational Investigation into the Human Representation and Processing of Visual Information*. San Francisco: W. H. Freeman.

Milkowski, M. (2016). Unification strategies in cognitive science. *Studies in Logic, Grammar and Rhetoric, 48*(1), 13–33.

Roth, M., & Cummins, R. (2017). Neuroscience, psychology, reduction, and functional analysis. In D. M. Caplan (Ed.), *Explanation and Integration in Mind and Brain Science* (pp. 29–43). New York: Oxford University Press.

Index

B

biolinguistics 32, 33, 36, 37
biologism 31–33
brain imaging 70, 73, 85
brain lesions 73, 76
Broca's aphasia 73, 74

C

canalization 15
categorical invariance 152
Cognitive Grammar 34, 108, 148, 218
competence grammar 226
complex predicates 37, 142, 184, 185, 187, 189–199, 201, 202
conceptual schemas 34, 147, 148
Conceptual Semantics 110, 192
conceptual spaces 113, 114, 172

Conservativity Constraint 163–165, 167, 169, 170, 174, 175, 177, 179, 181, 183
Construction Grammar 13, 34. *See also* Sign-Based Construction Grammar
content integration 151–154, 156, 158
critical period 51–53

D

double dissociation 78–80, 82

E

epistemic systematization 106
event linking 200, 201
event scanning 201

F
FOXP2 62, 63
functionalism 30–32, 34. *See also* multiple realizability

G
Generative Grammar 4, 8, 217, 218
grammar 3, 8, 71, 102, 204, 206, 216, 217, 220, 221, 225–228, 236
  derivational type vs. representational type 216, 218–221, 228

I
indexing 102, 209–215
intensionality 224
intra-event mapping 199–202
inverse problem 68, 69

L
late closure 24, 25
levels of organization 25, 44, 234. *See also* Marr's three-level schema
light verbs 185, 187–190, 193–195, 201, 218
linguistic relativity hypothesis 115, 119

M
Marr's three-level schema 44, 88
mental spaces 171
minimal attachment 24, 25

modularity 80
monotonicity 168
multiple realizability 30, 31

N
neurolinguistics 7, 29, 36, 78, 82, 83

P
parsing 102, 225–228
performance 22, 69, 74, 77, 160, 226
private speech 107
protolanguage 107

Q
quantifiers 159, 160, 163–165, 167–170, 172–175, 177–181, 183

R
reduction 26, 54, 83, 88, 137
representational linking 151–154, 156–158
representation invariance 222

S
savant studies 80, 82
sequencing 64, 187, 208, 209, 214, 215
serial verb constructions 185–187, 190–193, 195–201
Sign-Based Construction Grammar 13

specific language impairment (SLI) 62, 64, 79

T

topicalization 207, 214
Turing machine 45
twin studies 54, 56

U

unification 3, 4, 33, 36, 88, 154, 234

V

variable binding 37, 142–144, 146–153, 155–158
visual cognition 85

W

Wernicke's aphasia 73–75
Witness Set Constraint 168–170, 175, 183, 184
word order 37, 84, 142, 202–213, 215